Wireless Sensor Networks

Wireless Sensor Networks
Deployment Strategies for Outdoor Monitoring

Fadi Al-Turjman

CRC Press
Taylor & Francis Group
Boca Raton London New York

CRC Press is an imprint of the
Taylor & Francis Group, an **informa** business

MATLAB® is a trademark of The MathWorks, Inc. and is used with permission. The MathWorks does not warrant the accuracy of the text or exercises in this book. This book's use or discussion of MATLAB® software or related products does not constitute endorsement or sponsorship by The MathWorks of a particular pedagogical approach or particular use of the MATLAB® software.

CRC Press
Taylor & Francis Group
6000 Broken Sound Parkway NW, Suite 300
Boca Raton, FL 33487-2742

© 2018 by Taylor & Francis Group, LLC
CRC Press is an imprint of Taylor & Francis Group, an Informa business

No claim to original U.S. Government works

Printed on acid-free paper

International Standard Book Number-13: 978-0-8153-7581-4 (Hardback)

Library of Congress Cataloging-in-Publication Data

Names: Al-Turjman, Fadi, author.
Title: Wireless sensor networks: deployment strategies for outdoor monitoring / Fadi Al-Turjman.
Description: Boca Raton: Taylor & Francis, a CRC title, part of the Taylor & Francis imprint, a member of the Taylor & Francis Group, the academic division of T&F Informa, plc, [2017] | Includes bibliographical references.
Identifiers: LCCN 2017040978 | ISBN 9780815375814 (hb : acid-free paper) | ISBN 9781351207034 (e)
Subjects: LCSH: Wireless sensor networks. | Environmental monitoring.
Classification: LCC TK7872.D48 A3953 2017 | DDC 006.2/5--dc23
LC record available at https://lccn.loc.gov/2017040978

Visit the Taylor & Francis Web site at
http://www.taylorandfrancis.com

and the CRC Press Web site at
http://www.crcpress.com

This book is dedicated to my beloved parents, my mother and my father, to honor their major support and encouragement over the course of my entire life. In this way, I wish to express maybe a little part of my great gratitude to them.

My mother, Lama, thank you very much for the tender care you have been always providing. Your delicious morning sandwiches' taste in the cold winter is still in my mouth.

Great mother, words can never express the deepest gratitude I have for you. You have been there for me my whole life and I love you so much for it. I mostly admire and love you for the person YOU are.

My father, Ferzat, completed his BSc and MSc degrees in mathematics at Damascus University, Damascus, Syria, in 1974. He spent his entire life teaching not only math but also ethics. Throughout his career, Mr. Al-Turjman has made significant contributions in multiple areas of the education system. He received several academic honors for his outstanding teaching, including the Prince of Kuwait Prize for Excellence in Teaching twice, in 2000 and 2010. Indeed, he is a brilliant teacher and scientific communicator. I owe him a lot, and I believe his support was a key factor in any achievement I have ever made.

Thank you very much great father for your sincere love.

I love you always,

Your son,
Fadi Al-Turjman

Contents

Preface

Wireless Sensor Networks (WSNs) overcome the difficulties of other monitoring systems because they require no human attendance on site, provide real time interaction with events, and maintain efficient cost and power operations. However, further efficiencies are required especially in the case of Outdoor Environment Monitoring (OEM) applications due to their harsh operational conditions, huge targeted areas, limited energy budgets, and required three-dimensional (3-D) setups. A fundamental issue in defeating these practical challenges is the deployment planning of the WSNs. The deployment plan is a key factor of many intrinsic properties of OEM networks, summarized in connectivity, lifetime, fault-tolerance, and cost-effectiveness. In this book, we investigate the problem of WSNs' deployments which address these properties in order to overcome the unique challenges and circumstances in OEM applications.

MATLAB® is a registered trademark of The MathWorks, Inc. For product information, please contact:

The MathWorks, Inc.
3 Apple Hill Drive
Natick, MA, 01760-2098 USA
Tel: 508-647-7000
Fax: 508-647-7001
E-mail: info@mathworks.com
Web: www.mathworks.com

Author

Fadi Al-Turjman is an associate professor at Middle East Technical University Northern Cyprus Campus, Turkey. He received his PhD degree in computing science from Queen's University, Kingston, Ontario, Canada, in 2011. He is a leading authority in the areas of smart/cognitive, wireless, and mobile networks' architectures, protocols, deployments, and performance evaluation. His record spans more than 150 publications in journals, conferences, patents, books, and book chapters, in addition to numerous keynotes and plenary talks at flagship venues. He has received several recognitions and best papers' awards at top international conferences and led a number of international symposia and workshops in flagship ComSoc conferences. He is serving as the lead guest editor in several journals, including the *IET Wireless Sensor Systems, Sensors (Multidisciplinary Digital Publishing Institute), IEEE Access*, and *Wireless Communication and Mobile Computing* (Wiley). He is also the general workshops chair for the IEEE International Conference on Local Computer Networks (LCN '17). Recently, he has published his book entitled *Cognitive Sensors & IoT: Architecture, Deployment, and Data Delivery*

with Taylor & Francis Group, CRC Press, New York. Since 2007, he has been working on international wireless sensor networks projects related to remote monitoring, as well as Smart Cities-related deployments and data delivery protocols using integrated RFID–sensor networks.

1

INTRODUCTION

Wireless networking and advanced sensing technology have enabled the development of low-cost and power-efficient Wireless Sensor Networks (WSNs) which can be used in various domains such as health care, home intelligence, and Outdoor Environment Monitoring (OEM). Devices in WSNs, called Sensor Nodes (SNs), are used to sense certain properties of the surrounding environment, including physical/chemical properties, and transmit the sensed data to a central unit, called a Base Station (BS), either periodically or on demand. According to different application requirements, a WSN may consist of just a few or as many as thousands of wireless nodes, operating in a collaborative and coherent manner for a few days or several years to fulfill a specific task [1].

Currently, WSNs are proving to be a promising wireless monitoring technology in various OEM applications. In OEM applications, sensor nodes are assumed to monitor and report the status of the surrounding outdoor environments, as in air pollution, forest fire, and flood detection applications [2,3]. The wide use of WSNs in OEM is due to their enormous potential benefits and advantages. The direct and continuous existence of WSNs surrounding the monitored phenomena allows them to provide localized measurements and detailed information that is hard to obtain through traditional instrumentation (i.e., traditional sensors). For example, the Moderate Resolution Imaging Spectroradiometer (MODIS) [4,5], a traditional instrument, was launched to provide forest fire surveillance by capturing satellite images and processing them every 1–2 days, which is too late for the forest to survive. In contrast, in Reference [2], a WSN is used to provide real time interaction with forest fires due to the cooperative efforts of sensor nodes. Another example is the redwood tree application in Reference [6]. The entire surrounding of the redwood tree is known to have substantial variations in temperature and moisture along its length (~70 m). In order to study these variations, a winch near the top is needed to carry a set of weather monitoring instruments with a very long serial cable connected to a battery powered data logger at the tree base. However, a few small nodes constructing the WSN were able to collect the targeted data wirelessly at different scales and resolutions.

Measuring natural species with a WSN can also enable long term data collection at scales and resolutions that are difficult, if not impossible, to obtain otherwise. Moreover, due to the local computing and networking capabilities of WSNs, sensor nodes can be reprogrammed to do different tasks after deployment on site, based on changes in the conditions of the WSN itself and the targeted phenomena [1]. Therefore, laboratories can be extended to the monitored site where no human intervention is required, and real time interaction can be provided remotely.

Nonetheless, deploying* WSNs in OEM is challenging and needs further investigation due to harsh operational conditions, vast monitored areas, limited energy budgets, and required 3-D setups. For instance, most of the proposed OEM applications [2,3,6–9] are applied in harsh and large-scale areas, such as forests and river surfaces, and experience bad channel conditions as well as great chances of node failures and damages. Furthermore, some applications required more complicated deployment planning when 3-D setups [3,6,8,10], and long lifetime periods [2,6,9] are required. Therefore, we investigate efficient deployment plans to address these challenges and fill the literature gap in this critical research direction. In the following subsections, OEM deployment challenges and motivations behind this work are detailed. And major research contributions are reviewed in addition to outlining the remaining contents of this book.

1.1 Contributions

This work aims primarily at optimizing the WSN deployment in OEM applications by exploiting static/mobile RNs in grid-based deployments. We study the optimization problem when SNs' positions are known in advance and static RNs are utilized. Afterward, we

---

* Deployment is the process of identifying numbers and positions of the utilized nodes (devices) in the WSN.

examine a more complicated scenario where SNs' positions are anonymous and a hybrid of static and mobile RNs are in use. Accordingly, we present novel deployment schemes (strategies) to place the set of RNs/SNs on any grid model in these scenarios. The main contributions of this book are summarized as follows:

1. Generally, SNs/RNs can be placed on any vertex of the 3-D grid which results in a huge search space in large-scale applications, such as an OEM. Thus, this leads to extreme complexity in repositioning extra static or mobile RNs during the run time. We overcome this challenge by finding a subset of a relatively small number of grid vertices for these RNs without affecting the solution optimality.

2. Assuming prepositioned SNs and static RNs only, we consider them along with their communication links as a graph to mathematically represent their connectivity property. And thus, we formulate a solid mathematical optimization problem that has the aforementioned limited search space and aims at maximizing the network connectivity while maintaining cost and lifetime constraints.

3. Key to our contribution is also the use of proper cost and communication models in addition to a revised definition for the network lifetime, which is more appropriate in OEM applications.

4. Assuming the mobility feature in a subset of the available RNs, we also prolong the network lifetime. We mathematically formulate this problem as an Integer Linear Program (ILP) and provide a two-phase solution for it. We consider an effective power level metric at the objective function of the ILP, which is the minimum node residual energy along with the total consumed energy. This consideration results in more influential movement for the mobile RNs, and thus more equilibrium in the traffic load is achieved.

5. We derive an upper bound for the maximum network lifetime under idealistic operational conditions in OEM. The proposed two-phase solution shows significant improvements in terms of the network lifetime in comparison to this upper bound.

6. We investigate mobile vs. static grid-based deployments in order to overcome harsh operational conditions in OEM applications while considering hexagon and square grid shapes for these deployment solutions.

7. Genetic-based solutions, as well as solid mathematical optimization, have been exploited in significant OEM applications such as the emerging Smart City project.

In summary, this book provides guidelines for device deployment of typical heterogeneous WSNs in OEM applications.

1.2 Book Outline

The rest of this book is organized as follows. Chapter 2 highlights relevant background and related work in the literature. In Chapter 3, we introduce a novel 3-D deployment strategy, called Optimized 3-D Grid deployment with Lifetime Constraint (O3DwLC), for relay nodes in environmental applications. The strategy optimizes network connectivity while guaranteeing specific network lifetime and limited cost. Key to our contribution is a very limited search space for the optimization problem in addition to a revised definition for network lifetime, which is more appropriate in environment monitoring. The effectiveness of our strategy is validated through extensive simulations and comparisons, assuming practical considerations of signal propagation and connectivity. In Chapter 4, we propose two Optimized Relay Placement (ORP) strategies with the objective of federating disjoint WSN sectors with the maximum connectivity under a cost constraint on the total number of RNs to be deployed. The performance of the proposed approach is validated and assessed through extensive simulations and comparisons assuming practical considerations in outdoor environments. In Chapter 5, we propose a 3-D grid-based deployment for relay nodes in which the relays are efficiently placed on grid vertices. We present a novel approach, named Fixing Augmented network Damage Intelligently (FADI), based on a minimum spanning tree construction to reconnect the disjointed WSN sectors. The performance of the proposed approach is validated and assessed through extensive simulations; comparisons with two mainstream approaches are presented. Our protocol outperforms the related work in terms of the average relay node count and distribution, the scalability of the federated WSNs in large scale applications, and the robustness of the topologies formed. We elaborate further on this strategy in Chapter 6 where the hexagonal virtual grid is assumed instead of the square one. We propose the use of cognitive nodes (CNs) in the underlying sensor network to provide intelligent information processing and knowledge-based services to the end users. We identify tools and techniques to implement the cognitive functionality and formulate a strategy for the deployment of CNs in the underlying sensor network to ensure a high probability of successful data reception among communicating nodes. From MATLAB® simulations, we were able to verify that in a network with randomly deployed sensor nodes, CNs can be strategically deployed at predetermined positions, to deliver application-aware data that satisfies the end user's quality of information requirements, even at high application payloads. Chapter 7 proposes a 3-D grid-based deployment for heterogeneous WSNs (consisting of SNs, RNs, and MRNs). The problem is cast as a Mixed Integer Linear Program (MILP) optimization problem with the objective of maximizing the network lifetime while maintaining certain levels of fault-tolerance and cost efficiency. Moreover, an Upper Bound (UB) on the deployed WSN lifetime, given that there are no unexpected node/link failures, has been driven. Based on practical/harsh experimental settings in OEM, intensive simulations show that the proposed grid-based deployment scheme can achieve an average of 97% of the expected UB.

Additionally, a typical scenario has been discussed and analyzed in smart cities while considering genetic based approaches in Chapter 8. In this chapter, we study the path planning problem for these DCs while optimizing their counts and their total traveled distances. As the total collected load on a given DC route cannot exceed its storage capacity, it is important to decide on the size of the exchanged data packets (images, videos, etc.) and the sequence of the targeted data sources to be visited. We propose a hybrid heuristic approach for public data delivery in smart city settings. In this approach, public vehicles are utilized as DCs that read/collect data from numerously distributed Access Points (APs) and relay it back to a central processing base station in the city. We also introduce a cost-based fitness function for DCs election in the smart city paradigm. Our cost-based function considers resource limitations in terms of DCs' count, storage capacity and energy consumption. Extensive simulations are performed; results confirm the effectiveness of the proposed approach in comparison to other heuristic approaches with respect to total traveled distances and overall time complexity. Finally, Chapter 9 concludes the book and provides directions for future work.

References

1. I. Akyildiz, W. Su, Y. Sankarasubramaniam, and E. Cayirci, A survey on sensor networks, *IEEE Communications Magazine*, vol. 40, no. 8, pp. 102–114, 2002.
2. M. Hefeeda, Forest fire modeling and early detection uses wireless sensor networks, *School of Computing Science, Simon Fraser University*, Vancouver, BC, Technical Report TR-2007-08, August 2007.
3. F. Al-Turjman, Information-centric sensor networks for cognitive IoT: An overview, *Annals of Telecommunications*, vol. 72, no. 1, pp. 3–18, 2017.
4. G. Guo and M. Zhou, Using MODIS land surface temperature to evaluate forest fire risk of Northeast China, *IEEE Geoscience and Remote Sensing Letters*, vol. 1, no. 2, pp. 98–100, 2004.
5. NASA, MODIS—Moderate Resolution Imaging Spectroradiometer. http://modis.gsfc.nasa.gov/, 2009.
6. G. Tolle, J. Polastre, R. Szewczyk, and D. Culler, A macroscope in the redwoods, In *Proceedings of the ACM Conference on Embedded Networked Sensor Systems (SenSys)*, San Diego, CA, 2005, pp. 51–63.
7. J. Cox, MIT's tree-powered wireless network, *Network World*. www.networkworld.com/community/node/33278, 2009.
8. A. Singh, M. Batalin, V. Chen, M. Stealey, B. Jordan, J. Fisher, T. Harmon, M. Hansen, and W. Kaiser, Autonomous robotic sensing experiments at San Joaquin river. In *Proceedings of the IEEE International Conference on Robotics and Automation (ICRA)*, Rome, Italy, 2007, pp. 4987–4993.
9. B. Son, Y. Her, and J. Kim, A design and implementation of forest-fires surveillance system based on wireless sensor networks for South Korea mountains, *International Journal of Computer Science and Network Security (IJCSNS)*, vol. 6, no. 9, pp. 124–130, 2006.
10. J. Thelen, D. Goense, and K. Langendoen, Radio wave propagation in potato fields, In *Proceedings of the Workshop on Wireless Network Measurements*, Riva del Garda, Italy, April 2005.

2

DEPLOYMENT OF WIRELESS SENSOR NETWORKS IN OUTDOOR ENVIRONMENT MONITORING

An Overview

Deployment planning is of the utmost importance in the context of Wireless Sensor Networks (WSNs) as it decides the available resources and their configuration for system setup. This in turn plays a key role in the network performance. A significant amount of research has been made to enhance the network performance during its operational time by optimizing data routing and Medium Access Control (MAC) protocols [1]. Nevertheless, even with the best suit of routing and MAC protocols, a designed sensor network cannot achieve the targeted performance level unless it has been appropriately configured in advance. For example, if the installed devices are insufficient or if there are architecture deficiencies due to an ineffective deployment plan, the connectivity between the deployed nodes and the system lifetime will be degraded, in addition to causing nonoperational networks in some critical situations like flood and fire detection [2,3].

Accordingly, recent efforts are moving toward enhancing the network performance by optimizing the network deployment planning. Several network properties, considered as deployment objectives and constraints, have led to a rich research field. Using the categorization criteria adopted in Reference [4], we classify the technical approaches of WSN deployment into two groups, random deployment and deterministic (grid-based) deployment. In random deployment, nodes are arbitrarily scattered and are managed in an ad hoc manner. In contrast, in the grid-based deployment, nodes are placed according to virtual grid vertices which leads to more efficiency in terms of the

7

targeted network properties. The shortcomings of these approaches will be pointed out where appropriate in the following sections.

2.1 Desired Network Properties in OEM

Due to aforementioned challenges and characteristics, we will discuss the most important properties in the Outdoor Environment Monitoring (OEM) deployment planning, namely connectivity, fault-tolerance, and network (system) lifetime.

2.1.1 Connectivity

Connectivity is defined as the ability of the deployed nodes to stay connected with the Base Station (BS) to satisfy the targeted application objectives [5]. As previously mentioned, nodes and communication links in OEM are subject to several risks. Therefore, the deployed network connectivity is essential to ensure that the sensed data is delivered correctly to the BS for processing. In the context of WSNs, connectivity could be presented as k-connectivity. k-connectivity has two different meanings, namely, k-path connectivity and k-link connectivity. k-path connectivity means that there are k independent paths between the deployed nodes and the BS, while the k-link connectivity means that each node is directly connected to k neighboring nodes. In fact, a WSN could be disconnected even if k-link connectivity is satisfied. However, when $k > 1$, the network can tolerate some node and link failures. At the same time, the higher degree of connectivity improves communication availability among the deployed nodes. In the literature, some OEM deployments, such as the one in Reference [6], support k-path connectivity while others support k-link connectivity for more reliable and longer lasting networks, as in References [7–9]. In fact, it may not be necessary to maintain k-connectivity between all nodes, but only among the nodes which form the network backbone, which we call irredundant (critical) nodes.

In general, connectivity problems can be repaired after their occurrence either by using extra redundant nodes or by utilizing the node mobility feature. Node redundancy [10] is used to overcome disconnected networks. Redundant nodes, which are not being used for communication or sensing, are turned off. When the network becomes disconnected, one or more of the redundant nodes is turned on to resume the network connectivity. In Reference [11], the lowest number of redundant nodes are added to a disconnected static network, so that the network remains connected. In addition, overlapping clusters of sensor nodes are used to enhance the network connectivity in Reference [12]. In Reference [13], a distributed recovery algorithm is developed to address k-link connectivity requirements, where k is equal to 1 and 2. The idea is to identify the least set of nodes that should be repositioned in order to reestablish a particular level of connectivity. A shortcoming of such techniques is the requirement for extra nodes regardless of their roles, types, and physical positions, which may not be cost-effective in environment monitoring applications. Also, when some redundant nodes fail, it may no longer be possible to repair the network connectivity.

Meanwhile, node mobility can also be used to maintain network connectivity. Typically, mobile nodes are relocated after deployment to carry data between disconnected partitions of the network [14]. Providing radio connectivity using mobile nodes while considering ongoing missions, traveling distance, and message exchange complexity has also been considered recently in Reference [13]. However, relocating nodes without considering grid connectivity properties and characteristics can have severe effects on the direction of movement and the choice of the most appropriate node to be moved.

Although the above techniques can aid with repairing connectivity problems, they do not address the sources of these problems. In this book, we present a radically different approach toward addressing connectivity problems and complementing the work of the aforementioned techniques. We address the properties of grid connectivity in practice and under realistic scenarios, such as inaccurate positioning and communications irregularity. Thus, more efficient connectivity planning and maintenance can be achieved.

2.1.2 Fault-Tolerance

Due to the harsh operational conditions where OEM sensor networks are deployed, WSNs are susceptible to unpredictable events such as hardware failures, power leakage, physical damage, environmental interference, etc. Thus, it is crucial to ensure that WSNs are also fault-tolerant. As mentioned before, the sources of failure in OEM can occur because of either node or link failures. These failures are characterized by the Probability of Node Failure (PNF) and Probability of Disconnected Nodes (PDN) to indicate the harshness level of the monitored environment.

Three scenarios are considered in the deployed networks to be fault-tolerant: fault detection, fault diagnosis, and fault recovery [15]. There are many techniques for fault detection and diagnosis in the literature; they are classified into self-diagnosis, group detection, and hierarchal detection [16]. On the other hand, few fault recovery techniques are addressed in the literature. They can be classified into node and link recovery techniques. These recovery techniques assure the ability of the deployed network to stay operational and fulfill the assigned task in the presence of a set of nonfunctional nodes/links, so that sensed and carried data can be recovered. Although both node and link failures are important to be tolerated together, some deployments have considered only node failure recovery such as [9], while others have considered only the link failure recovery as in Reference [17], which weakens the deployed network performance. Compared to the complexity of considering a single type of these recovery techniques, it is trivial to consider both of them in one deployment technique by employing redundant nodes and maximizing the communication connectivity.

2.1.3 Lifetime

As for the system lifetime, it is the most challenging problem in any OEM deployment as the utilized nodes are energy constrained, and deployed networks are sometimes required to work for longer intervals measured in years [18,19]. In addition, it is undesirable to revisit harsh environments, such as those targeted by the OEM applications, simply for node replacement or recharging. Accordingly, accurate lifetime prediction in the early deployment planning stages is required. There are two types of lifetime predictions in the literature: node and network lifetime predictions. Node lifetime can be measured in several ways. For example, it can be measured based on the number of cycles over which the data is collected, on the cumulative active time of the node before it is depleted, or on the cumulative traffic volume of the node before its energy is depleted. However, all of the afore-mentioned measuring methods require accurate energy consumption models which consider the effects of surrounding environments for more realistic and practical predictions. This issue is addressed in our research. In fact, nodes in WSNs consume power in three main domains: sensing, communication, and data processing. Because the majority of power consumed per node is used in communica-tion, related work in literature has concentrated on proposing energy consumption models for receiving and transmitting wireless signals. Network lifetime has several definitions in the literature. One of the most common states that, "it is the time until the first node death occurs" [20]. Such a definition may not be suitable if we are monitor-ing, for example, the forest temperature or humidity as long as we are still receiving the same information from other nodes in the same area. Similarly, the definition, which says that "it is the time until the last node death occurs", can serve as an upper bound of other definitions. Thus, both of these definitions are unrealistic for OEM applications. Another way to define lifetime relies on the percent-age of nodes alive [20]. But choosing such a percentage threshold is usually arbitrary in the literature and does not reflect the application requirements. In addition, being alive does not mean that the node is still connected to the sink node (or BS). Yet another approach to evaluate network lifetime is in terms of connectivity and network par-titions [21]. These definitions still do not satisfy OEM applications. For instance, if a set of nodes is destroyed because of the movement of an animal or a falling tree in the monitored site, the network can still be considered functional as long as some other nodes are still alive and connected to the BS. Accordingly, redundant nodes and links must be used in such applications. In order to take into consideration node/link redundancy, an OEM specific lifetime definition is proposed in this book.

2.2 Random vs. Deterministic WSNs Deployment

In the literature, deployments targeting network properties men-tioned above are mainly classified into two categories: random vs. deterministic (grid-based) deployments.

2.2.1 Random Deployment

In random deployments, nodes can be placed randomly to reduce the deployment cost, or based on a weighted random deployment planning, they can be placed where the distributed nodes' density is not uniform in the monitored areas. For instance, K. Xu et al. [22] studied the random RN deployment in a 2-D plane. The authors proposed an efficient network lifetime maximization deployment when the RNs are directly communicating with the BS. In this study, it was established that different energy consumption rates at different distances from the BS render uniform RN deployment, thus being a poor candidate for network lifetime extension. Alternatively, a weighted random deployment is proposed. In this random deployment, the density of RN deployment increases as the distance to the BS increases, and thus distant RNs can split the traffic among themselves. This in turn extends the average RN lifetime. We note that some attempts have been made toward the 3-D deployment as well. For example, authors in Reference [23] consider the implications of sensing and communication ranges on the network connectivity in random 3-D deployments. In Reference [24], the authors proposed a distributed algorithm that achieves k-connectivity in homogenous WSNs randomly deployed in the 3-D space. Simply, the idea is to adapt the nodes' transmission power until either the distance separating two consecutive neighbors is greater than a specific threshold or the maximal power is reached. However, this method is not cost-effective due to hardware complexity required to adapt the transmission ranges, which is not worthy in OEM applications. In addition, adapting the transmission range to reach farther distances would increase the energy consumption, and hence degrade the network lifetime.

2.2.2 Deterministic (Grid-Based) Deployment

Proposals described above are suitable for applications which are not interested in the exact node positioning. In contrast, some proposals have advocated deploying nodes exactly on specific predefined locations, called grid vertices. These locations are optimized in terms of the aforementioned network properties and the feasibility of the location itself in reality (e.g., non-reachable locations are not feasible for node deployment). Due to the interest posed by OEM applications in the exact physical positioning of sensor and relay nodes, this type of deployment (i.e., grid-based deployment) serves this purpose more appropriately, and hence is adopted in our work. Moreover, coupling this type of deployment with guaranteed multipath routing can significantly enhance the network lifetime and fault-tolerance chances, in addition to repairing network connectivity problems.

In the literature, two categories of approaches have been pursued to enhance these three main properties (i.e., network lifetime, connectivity, and fault-tolerance): (1) *populating additional (redundant) nodes*, and (2) *employing mobile nodes*. The approach presented in Reference [13] counters faulty sensor nodes by repositioning mobile pre-identified spare sensors from different parts of the network. The most appropriate candidate spare node is the closest one. In order to detect the closest spare to the faulty sensor, a grid-based approach is proposed. The targeted region is divided into cells. Each cell has a head advertising available spare nodes in its cell or requesting the spare ones for it. Once the spares are located, they are moved to the cell of failure without affecting the data traffic and the network topology. However, assuming that the closest node will always be moved to recover the faulty ones will rapidly deplete their batteries unless a load balance constraint is added.

W. Alsalih et al. [33] proposed a deterministic deployment for data gathering using a mobile node, called a data collector (DC). In this deployment, a mobile data collector moves along a predefined track through the sensing field. SNs whose transmission ranges overlap with this track are called relay nodes (RNs). In addition to forwarding their own measurements to the data collector, the RNs collect the measurements of neighboring SNs whose transmission ranges do not overlap with the track. It was shown that employing mobile data collectors can extend the network lifetime in comparison to conventional WSNs employing static nodes only. In fact, the concept of mobile nodes, or data collectors, was used earlier in References [25,26]. These references assumed the existence of a number of predefined locations where data collectors can be placed. The network lifetime was divided into equal length time periods, called rounds (a more rigorous definition of a round will be given subsequently in this book), and data collectors are moved to new locations at the beginning of each round. In Reference [25], the problem of finding the optimal locations for the mobile nodes was formulated as a Mixed Integer Linear Program (MILP) problem whose objective is minimizing the total consumed energy during a round. It was also shown that the identified optimal locations were optimal when the objective was to minimize the maximum energy spent by a single SN. However, these two objective functions are not suitable for the placement of mobile nodes since the optimal solutions will not change over the time, i.e., the maximum or total energy spent per round might not change; hence, locations of data collectors (mobile nodes) will not be changed. In contrast, a heuristic mobile node placement scheme that considers nodes' residual energy was proposed in Reference [26]. The locations of data collectors were chosen according to local information only. In other words, the decision of whether or not a mobile node is placed at a given location is made based on the residual energy of SNs that are one hop away from that location. Consequently, the locations calculated may be far from optimal.

Even so, assuming all nodes are able to move may not be valid in a number of OEM applications. It is more appropriate if a subset of the deployed nodes is assumed to have the mobility feature, or more static nodes are populated to repair connectivity (by recovering faulty nodes) and prolong the network lifetime. In an earlier work [27], we proposed an Integer Linear Program (ILP) that assumes a number of the deployed nodes have the ability to change their locations during the network operational time. This deployment strategy enforces the minimal energy consumption while maintaining connectivity requirements, fault-tolerant constraints, and cost-effectiveness. A key parameter in formulating the deployment problem is limiting the 3-D search space to a finite set of points by using grid models. The performance of the proposed strategy is validated through extensive simulations. Network lifetime enhancements of up to 50% have been achieved as compared to other deployment schemes under practical settings in OEM applications.

Unlike the previous references, the authors in Reference [28] proposed an algorithm to achieve fault-tolerant and long lasting WSNs by populating static nodes which have at least two disjointed paths between every pair of sensor nodes. This algorithm simply identifies candidate positions for relay nodes that cover the maximum number of sensors. Such candidate positions are found at the intersections of the communication ranges of neighboring sensor nodes. Relay nodes with the widest coverage span are then placed at these candidate positions. The algorithm checks whether the relays form a 2-connected graph and every sensor can reach at least two relays. If not, more relays are added and the connectivity and coverage are rechecked. This algorithm is repeated until the objective is achieved. However, relay positions may not be accurate in reality due to communication range irregularity. This work becomes difficult when 3-D space is considered. Furthermore, lifetime and energy constrains are not addressed in this algorithm, in addition to ignoring node recovery (i.e., ignoring the lost data by the faulty node) leading to link fault-tolerance deployment rather than link and node fault-tolerance. This makes the proposed deployment unsuitable for OEM applications. Similarly, static nodes are used in Reference [29] to repair connectivity in grid-based deployments. In this work, the authors aim at populating additional minimum numbers of relay nodes to establish connections with partitioned segments in the network using a heuristic approach. It shows beneficial aspects of the resulting topology with respect to connectivity and traffic balance. Nevertheless, node mobility, which may have a dramatic impact on both connectivity and lifetime of the network, hasn't been addressed in this work. Moreover, adding additional relay nodes during the monitoring process is not always applicable under the harsh operational conditions experienced by the OEM applications. Finally, this work is adopted in a 2-D plane only. Contrarily, in Reference [30], maximal 3-D connectivity using the least number of nodes has been discussed, and thus it provides a cost-effective deployment plan. However, this work assumes regular communication range represented by a binary sphere around each grid vertex, which is not the case in practice. In addition, there is no fault-tolerance and lifetime constraints in this work, making it unsuitable for OEM applications. We note that this work is restricted to a specific type of grid model, which is the octahedron 3-D grid, and thus it is not applicable on other 3-D grid shapes (e.g., cubes, pyramids, etc.).

2.3 Summary

In this section, we conclude our literature overview. Reviewed deployment proposals are classified and summarized in Table 2.1. In addition, corresponding proposals to reviewed deployments are compared based on the desired network properties in OEM applications. The notation (1) in Table 2.1 indicates that the technique of additional (redundant) nodes is used to achieve the desired network properties, while the notation (2) indicates that node mobility is used for that purpose. As for 2-D and/or 3-D, it means that the deployment is designed for two-dimensional and/or three-dimensional space, respectively. The *Network-based* and/or *Node-based* under the lifetime property indicate that the system lifetime is measured based on the overall network status (e.g., network partition) or on a single node status (e.g., first node death). Under fault-tolerance, link fault-tolerance means that the tolerated source of failure is the link, while node fault-tolerance means that the tolerated source of failure is the node. The 1-path under connectivity refers to providing at least one path from each node to the BS during the monitoring process. Finally, the "—" means that the network property is not addressed in the corresponding proposal.

Unsurprisingly, most of the proposed deployments, which focus on the network connectivity and lifetime and/or fault-tolerance, have addressed the cost factor of the deployed network in terms of total number of nodes used. This is due to high probabilities of node damage and loss in applications requiring these network properties. In contrast, none of the proposed deployments have addressed all of the required properties, except Reference [33]. Even in Reference [33], a 1-path connectivity property is addressed rather than k-path and/or k-link connectivity, where $k \geq 1$. Also, it is a random deployment technique. Moreover, lifetime and fault-tolerance properties in this reference are restricted to *node-based* and *node fault-tolerance* only, while it is necessary in OEM to consider the *network-based* and *link fault-tolerant* properties as well. Hence, there is no deployment proposal which considers all of the required network properties in OEM applications. Also, there is no proposal considering both link and node fault-tolerance simultaneously. Similarly, there is no proposal considering both node-based and network-based system lifetime simultaneously. Furthermore, it is obvious from Table 2.1 that

Table 2.1 A Comparison between Various Deployment Proposals in the Literature Considering the Desired Network Properties in OEM

REFERENCE	NETWORK PROPERTIES				DEPLOYMENT		TARGETED SPACE
	COST[a]	CONNECTIVITY	FAULT-TOLERANCE	LIFETIME	TYPE	TECHNIQUE	
[13]	√	1-path	—	Network-based	Grid-based	(2)	2-D
[31,30]	√	k-path	Link fault-tolerant	—	Grid-based	(1)	3-D
[32]	√	k-path	Link fault-tolerant	—	Grid-based	(1)	2-D
[24]	—	k-link	Link fault-tolerant	—	Random	(1)	3-D
[29]	√	1-path	—	Network-based	Grid-based	(1)	2-D
[23]	—	k-link	Link fault-tolerant	—	Random	(1)	3-D
[22]	√	1-path	—	—	Random	(1)	2-D
[33]	√	1-path	Node fault-tolerant	Node-based	Random	(1)	2-D
[27]	√	1-path	—	Node-based	Grid-based	(2)	2-D
[27]	√	k-path	Node-link fault-tolerant	Network-based	Grid-based	(2)	3-D

[a] The √ in this column indicates that the cost factor is considered in the corresponding reference.

there is no proposal considering the deployment technique of type (2) in the 3-D space. Although there are some proposals considering the deployment technique (2) in 2-D plane, they are not addressing the case where heterogeneous hybrid nodes are used (i.e., when a subset of the RNs is mobile). Consequently, we are in a dire need for generic 3-D grid-based deployments considering both techniques, (1) and (2), to achieve the desired OEM network properties. We need a deployment that addresses technique (2) where hybrid nodes are utilized. In addition, we need a deployment that addresses technique (1) such that the additional nodes are considered and deployed at the beginning of the monitoring process. Moreover, a deployment plan considering both node- and network-based lifetime definitions, in addition to communication links irregularity in 3-D space, is required.

References

1. I. Akyildiz, W. Su, Y. Sankarasubramaniam, and E. Cayirci, A survey on sensor networks, *IEEE Communications Magazine*, vol. 40, no. 8, pp. 102–114, 2002.
2. M. Hefeeda, Forest fire modeling and early detection uses wireless sensor networks, *School of Computing Science, Simon Fraser University*, Vancouver, BC, Technical Report TR-2007-08, August 2007.
3. F. Al-Turjman, Information-centric sensor networks for cognitive IoT: An overview, *Annals of Telecommunications*, vol. 72, no. 1, pp. 3–18, 2017.
4. M. Younis and K. Akkaya, Strategies and techniques for node placement in wireless sensor networks: A survey, *Elsevier Ad Hoc Network Journal*, vol. 6, no. 4, pp. 621–655, 2008.
5. M. Ibnkahla, Ed., *Adaptation and Cross Layer Design in Wireless Networks*, CRC Press, Boca Raton, 2008.
6. G. Tolle, J. Polastre, R. Szewczyk, and D. Culler, A macroscope in the redwoods, In *Proceedings of the ACM Conference on Embedded Networked Sensor Systems (SenSys)*, San Diego, CA, 2005, pp. 51–63.
7. M. Hamilton, E. Graham, P. Rundel, M. Allen, W. Kaiser, M. Hansen, and D. Estrin, New approaches in embedded networked sensing for terrestrial ecological observatories, *Environmental Engineering Science*, vol. 24, no. 2, pp. 192–204, 2007.
8. R. Pon, A. Kansal, D. Liu, M. Rahimi, L. Shirachi, W. Kaiser, G. Pottie, M. Srivastava, G. Sukhatme, and D. Estrin, Networked infomechanical systems (NIMS): Next generation sensor networks for environmental monitoring, *IEEE MTT-S International Microwave Symposium Digest*, vol. 1, pp. 373–376, 2005.

9. N. Ramanathan, L. Balzano, M. Burt, D. Estrin, E. Kohler, T. Harmon, C. Harvey, J. Jay, S. Rothenberg, and M. Srivastava, Rapid deployment with confidence: Calibration and fault detection in environmental sensor networks, *Center of Embedded Network Systems*, Los Angeles, CA. www. scholarship.org/uc/item/8v26b5qh, 2006.
10. A. Cerpa and D. Estrin, Ascent: Adaptive self-configuring sensor networks topologies, *IEEE Transactions on Mobile Computing*, vol. 3, no. 3, pp. 272–285, 2004.
11. N. Li and J. C. Hou, Improving connectivity of wireless ad hoc networks. In *Proceedings of the IEEE International Conference on Mobile and Ubiquitous Systems: Networking and Services (MobiQuitous)*, San Diego, CA, 2005, pp. 314–324.
12. M. Youssef, A. Youssef, and M. Younis, Overlapping multihop clustering for wireless sensor networks, *IEEE Transactions on Parallel Distributed Systems*, vol. 20, no. 12, pp. 1844–1856, 2009.
13. A. Abbasi, U. Baroudi, M. Younis, and K. Akkaya, C2AM: An algorithm for application-aware movement-assisted recovery in wireless sensor and actor networks, In *Proceedings of the ACM International Wireless Communications and Mobile Computing Conference (IWCMC)*, Leipzig, Germany, 2009, pp. 655–659.
14. M. Dunbabin, P. Corke, I. Vasilescu, and D. Rus, Data muling over underwater wireless sensor networks using an autonomous underwater vehicle, In *Proceedings of the IEEE International Conference on Robotics and Automation (ICRA)*, Orlando, FL, 2006, pp. 2091–2098.
15. M. Yu, H. Mokhtar, and M. Merabti, A survey on fault management in wireless sensor networks, *ACM Journal of Network and Systems Management*, vol. 15, no. 2, pp. 171–190, 2007.
16. L. Moreira, H. Vogt, and M. Beigl, A survey on fault tolerance in wireless sensor networks, *Scientific Commons*. http://en.scientificcommons. org/23851530, 2009.
17. J. Thelen, D. Goense, and K. Langendoen, Radio wave propagation in potato fields, In *Proceedings of the Workshop on Wireless Network Measurements*, Riva del Garda, Italy, April 2005.
18. A. Mainwaring, J. Polastre, R. Szewczyk, D. Culler, and J. Anderson, Wireless sensor networks for habitat monitoring, In *Proceedings of the ACM International Workshop on Wireless Sensor Networks and Applications (WSNA)*, Atlanta, GA, 2002, pp. 399–423.
19. P. Juang, H. Oki, Y. Wang, M. Martonosi, L. Peh, and D. Rubenstein, Energy-efficient computing for wildlife tracking: Design trade-offs and early experiences with ZebraNet, *ACM SIGPLAN Notices*, vol. 37, no. 10, pp. 96–107, 2002.
20. L. van Hoesel, T. Nieberg, J. Wu, and P. Havinga, Prolonging the lifetime of wireless sensor networks by cross layer interaction, *IEEE Wireless Communications*, vol. 11, no. 6, pp. 78–86, 2004.
21. Y. Chen and Q. Zhao, On the lifetime of wireless sensor networks, *IEEE Communications Letters*, vol. 9, no. 11, pp. 976–978, 2005.

22. K. Xu, H. Hassanein, G. Takahara, and Q. Wang, Relay node deployment strategies in heterogeneous wireless sensor networks, vol. 9, no. 2, pp. 145–159, 2010.

23. V. Ravelomanana, Extremal properties of three-dimensional sensor networks with applications, *IEEE Transactions on Mobile Computing*, vol. 3, no. 3, pp. 246–257, 2004.

24. M. Ishizuka and M. Aida, Performance study of node placement in sensor networks, In *Proceedings of the International Conference on Distributed Computing Systems Workshops (ICDCSW)*, Tokyo, Japan, 2004, pp. 598–603.

25. S. Gandham, M. Dawande, R. Prakash, and S. Venkatesan, Energy efficient schemes for wireless sensor networks with multiple mobile base stations, In *Proceedings of the IEEE Global Telecommunications Conference (GLOBECOM)*, San Francisco, CA, 2003, pp. 377–381.

26. A. Azad and A. Chockalingam, Mobile base stations placement and energy aware routing in wireless sensor networks, In *IEEE Wireless Communications and Networking Conference (WCNC)*, Las Vegas, NV, 2006, pp. 264–269.

27. F. Al-Turjman, H. Hassanein, and M. Ibnkahla, Towards prolonged lifetime for deployed WSNs in outdoor environment monitoring, *Elsevier Ad Hoc Networks Journal*, vol. 24, no. A, pp. 172–185, Jan 2015.

28. B. Hao, H. Tang, and G. Xue, Fault-tolerant relay node placement in wireless sensor networks: Formulation and approximation, In *Proceedings of the Workshop on High Performance Switching and Routing (HPSR)*, Phoenix, AZ, April 2004.

29. S. Lee and M. Younis, Optimized relay placement to federate segments in wireless sensor networks, *IEEE Journal on Selected Areas in Communications*, vol. 28, no. 5, pp. 742–752, 2010.

30. S. N. Alam and Z. Haas, Coverage and connectivity in three-dimensional networks, In *Proceedings of the ACM International Conference on Mobile Computing and Networking (MobiCom)*, Los Angeles, CA, 2006, pp. 346–357.

31. K. Xu, H. Hassanein, G. Takahara, and Q. Wang, Relay node deployment strategies in heterogeneous wireless sensor networks: Multiple-hop communication case, In *Proceedings of the IEEE Conference on Sensor and Ad Hoc Communications and Networks (SECON)*, Santa Clara, CA, 2005, pp. 575–585.

32. F. Al-Turjman, M. Ibnkahla, and H. Hassanein, An overview of wireless sensor networks for ecology and forest monitoring, In *Proceedings of the IEEE International Workshop on Signal Processing and its Applications (WoSPA)*, Sharjah, UAE, 2008.

33. W. Alsalih, H. Hassanien, and S. Akl, Placement of multiple mobile data collectors in underwater acoustic sensor networks, *ACM Journal of Wireless Communications and Mobile Computing*, vol. 8, no. 8, pp. 1011–1022, 2008.

3

EFFICIENT DEPLOYMENT OF WIRELESS SENSOR NETWORKS TARGETING ENVIRONMENT MONITORING APPLICATIONS*

Wireless Sensor Networks (WSNs) enable long term environment monitoring at scales and resolutions that are difficult, if not impossible, to obtain using conventional techniques. WSNs can be re-tasked after deployment in the field based on changes in the environment, conditions of the sensor network themselves, or scientific endeavor requirements [3]. However, most environmental applications require all seasons data-gathering which can include extreme conditions. Thus, long-lasting, tightly connected networks in the monitored field are needed in order to have better understanding of the monitored phenomena such as the life cycles of huge redwood trees [34] or to satisfy specific application objectives for several years as in forestry fire detection [31].

One of the most efficient techniques used for maintaining long-lasting connected networks is achieved by deploying Relay Nodes (RNs) [19]. Relay nodes can have extra storage space and much more powerful transceivers to forward sensed data for long distances in huge monitored sites; thus, the energy of Sensor Nodes (SNs) is saved for further data sensing and gathering. Nevertheless, deployment of relay nodes [11,24] in environmental applications is a challenging problem due to harsh environments and required 3-D setups.

Harshness of the environment arises because of the nature of outdoor monitoring applications where sensor networks may work

---

* This chapter has been coauthored with Hossam S. Hassanein and Mohamed A. Ibnkahla.

under heavy rain, extreme temperature variations, and sometimes stormy days destroying the deployed nodes and/or their communication links (edges). Many nodes and communication links may also be destroyed by unexpected visitors such as birds and wild fauna. In addition, dense trees and growing foliage can weaken communication links and affect connectivity. Nodes and links are prone to risks which lead to high probabilities of failures. Many nodes may become disconnected, which can degrade the overall network lifetime. As a result, we are characterizing harshness of the monitored environment by the Probability of Node Failure (PNF) and Probability of Disconnected Nodes (PDN). Due to high PNF and PDN in environmental applications, it is reasonable to have redundant nodes in the deployed sensor networks [23]. In the scope of this chapter, a *redundant* node is the node which can be removed from the network without affecting the targeted data. Contrarily, an *irredundant* node is defined as a unique source of information in the monitored site that cannot be recovered by other nodes in the network.

Meanwhile, deployment in environmental applications becomes more challenging when 3-D setups are required. In environmental applications, relay nodes are not only forwarding data from different variations in the horizontal plane but also from different vertical levels (e.g., on trees, at soil surface, and even underground). For instance, in monitoring the massive redwood trees in California, some experiments required sensor placements at different heights on the trees spanning a range of tens of meters [34]. There is also increased interest in 3-D environmental applications such as CO_2 flux monitoring and imagery [31], where sensors are placed at different vertical levels to fulfill coverage and data accuracy requirements. Accordingly, communication links between deployed sensor nodes have to be considered in a 3-D space rather than a 2-D plane only, raising the complexity of the deployed network connectivity [12,26].

Nonetheless, existing deployment schemes for such applications [28,29,31] are not always based on sound connectivity models but rather on simplified lifetime models. They have not efficiently addressed the problem of connectivity in a 3-D space, which is a natural model in environmental applications. For example, deployment in Reference [18] focused on connectivity in 2-D outdoor applications. Deployment algorithms are proposed in References [25,33] to guarantee the 2-D connectivity and/or ensure survivability in case of node failure without lifetime considerations. In contrast, the deployment in Reference [15] is aimed at maximizing the network lifetime under specific energy budget. The energy provisioning and relay node placement are formulated as a mixed-integer linear programming problem in a 2-D plane. Heuristic algorithms are then introduced to overcome the computational complexity. Nevertheless, relay nodes in Reference [15] are assumed to adapt their transmission range to reach any other node in the network, and connectivity is not considered as an issue (which is not practical in environmental applications). A hybrid approach has been proposed in Reference [35] to balance connectivity and lifetime in 2-D outdoor deployments. Even in the most recent papers in environmental WSN applications, including volcano [30] and harsh industrial [21] environment monitoring, network connectivity is considered in a 2-D plane with the assumption of a very basic binary communication disc model which is not realistic. Thus, these approaches are prone to failure in practical large-scale environmental applications.

It is worth mentioning that there are some attempts toward the 3-D deployment. As an example, authors in Reference [18] studied the effects of sensing and communication ranges on connectivity in a 3-D space. However, connectivity optimization with lifetime constraints has not been investigated so far in 3-D environmental applications. As noted in Reference [37], many of the popular deployment strategies, which are optimally solved in polynomial time in a 2-D plane, become NP-Hard in 3-D settings.

We also note that relay nodes in environment monitoring are generally more expensive than sensing nodes due to the cost of the wide range transceivers used to cover large-scale areas, such as forests and cities. Therefore, *efficient* 3-D deployment must maintain connectivity and lifetime that limits the number of these expensive nodes. In *efficient* environmental deployments, it is also undesirable to revisit the monitored sites (e.g., for node replacement or battery recharging). Therefore, the deployed wireless sensor network must be guaranteed to function for a pre-specified lifetime period. For more accurate and practical lifetime guarantees, an environment-specific lifetime definition should be considered. Even though there are several definitions for WSN lifetime in the literature, there is no agreement on a definition for lifetime in environmental applications. An inappropriate definition might lead to incorrect lifetime estimation, and hence cause a waste of resources.

In this chapter, we investigate an efficient way for the relays' placement that address aforementioned challenges and desired WSNs' features in environmental applications. Such node placement problem has been shown in Reference [5] to be NP-hard. Finding nonoptimal approximate solutions is also NP-hard in some cases [13]. To address this complexity, we propose an efficient two-phase relay node deployment in a 3-D space, called O3DwLC strategy. The first phase of O3DwLC is used to set up a connected network backbone using minimum number of relay nodes for cost efficiency. In the second phase, we aim at finding a set of a relatively small number of candidate positions. We do this by optimizing the relay node placement on these positions to achieve the maximum backbone connectivity for guaranteed lifetime period and within a limited cost budget. This two-phase deployment scheme will provide a reliable interaction with the network end users monitoring outdoor environments. It will optimize the relay node deployment in forests to detect fires and report wildlife activities in water bodies to record events concerning floods, water pollution, coral reef conditions, and oil spills. In addition, it will target other rural and hazardous areas such as deserts, polar and volcanic terrains, and battlefields.

Major contributions of this chapter are listed as follows. We explore the most suitable lifetime definition in environment monitoring. The appropriateness of the proposed definition is evaluated and compared to other lifetime definitions in the literature based on harsh environmental characteristics. We introduce a generic 3-D relay node placement problem which aims at maximizing connectivity with constraints on wireless sensor network cost and lifetime. We propose an efficient two-phase solution for the 3-D deployment problem, which considers a limited search space, generic communication model, most appropriate lifetime definition, and harsh operational conditions. Performance of the proposed two-phase solution is evaluated and compared to currently used strategies in environmental applications in the presence of varying probabilities of node failure and disconnection.

The remainder of this chapter is organized as follows. In Section 3.1, related work is outlined. Practical system models and placement problems are presented in Section 3.2. In Section 3.3, our two-phase deployment strategy is described. The performance of the proposed strategy is evaluated and compared to other deployment strategies in Section 3.5. Finally, conclusions and future work are given in Section 3.5.

3.1 Related Work

Extensive work has been reported in the literature relating to relay node deployment strategies which are classified into random vs. grid-based deployments [37]. In random deployment, nodes are randomly scattered and are organized in an ad hoc manner. In grid-based deployment, nodes are placed on grid vertices leading to more accurate positioning and data measurements. In addition, applying grid deployments in a 3-D space has several other benefits. Such deployments precisely limit the search space of relay node positions and possible paths between them. Thus, the formulation of the placement optimization problem is simplified. Moreover, grid models can reflect channel conditions using specific signal propagation models that consider harsh environment characteristics; hence, better connectivity can be maintained. Due to the interest of environmental applications in the exact physical positioning of sensor and relay nodes, grid-based deployment is the most appropriate option and is adopted for this work. However, more efficiency is required in grid-based deployments to enhance deployed node connectivity and limit the huge number of candidate positions (or grid vertices) in large-scale environmental applications.

Connectivity of the deployed grid-based network could be presented as k-connectivity. k-connectivity has two different meanings, namely, k-path connectivity and k-link connectivity [17]. The k-path connectivity means that there are k independent paths between every pair of nodes, while k-link connectivity means that each node is directly connected to k neighboring nodes. Nevertheless, a wireless sensor network could be disconnected even if k-link connectivity is satisfied. With k-path connectivity where $k \geq 1$, the network can tolerate some node and link failures. At the same time, the higher degree connectivity improves communication capacity among nodes. In some cases, it may not be necessary to maintain k-connectivity among all of the network nodes, but only among nodes which form the communication backbone of the network, and thus is adopted in this chapter. Nodes constructing the network backbone are called *irredundant* (critical) nodes in this research. Failure of connectivity between these nodes may lead to severe effects on the WSNs' performance such as network partitioning and data loss. Accordingly, node redundancy in Reference [8] is used to overcome such connectivity problems. Redundant nodes are deployed; the ones not being used for communication or sensing are turned off. When the network becomes disconnected, one or more of the redundant nodes is turned on to repair connectivity. In Reference [22], the lowest number of redundant nodes is added to a disconnected static network, so the network remains connected. Similarly, authors in Reference [20] focus on designing an optimized approach for connecting disjointed WSNs segments by populating the least number of relays. The deployment area is modeled as a grid with equal-sized cells. The optimization problem is then mapped by selecting the fewest count of cells to populate relay nodes such that all segments are connected. Overlapping clusters of sensor nodes, which rely on the concept of redundancy as well, are then used to enhance the network connectivity in Reference [37]. In Reference [1], a distributed recovery algorithm is developed to address 1- and 2-connectivity requirements. The idea is to identify the least set of nodes that should be repositioned in order to reestablish a particular level of connectivity. Nonetheless, redundant node deployment becomes an intricate task in huge 3-D spaces, where numerous positioning options are possible with different connectivity levels (degrees). Therefore, more efforts are required to optimize the deployment process under such circumstances.

Meanwhile, grid-based deployment should not only guarantee connectivity in harsh environment monitoring but also should guarantee specific network lifetime. Lifetime has several definitions in the literature. One of the most common lifetime definitions states "it is the time till the first node death occurs" [15]. Such a definition may not be appropriate if we are monitoring forest temperature or humidity because if a node dies, we can still receive similar (or redundant) information from other nodes in the same area. Therefore, this definition may only be considered as a lower boundary of other lifetime definitions and should not reflect the actual network lifetime.

Similarly, the definition, "it is the time till the last node death occurs," can serve as an upper bound of other definitions. Both of these definitions are unrealistic for environmental applications. Another way to define lifetime could rely on the percentage of alive nodes (which have enough energy to accomplish their assigned tasks) [15]. But choosing a percentage threshold is usually arbitrary and does not reflect the application requirements. In addition, being alive does not mean that the node is still connected. Therefore, other approaches to evaluate network lifetime are relying on connectivity and network partitions [9]. Still these definitions do not satisfy the environment monitoring applications. For instance, if a set of nodes is destroyed because of the movement of wild animals or falling trees in the monitored site, the network can still be considered functional as long as other nodes are still alive and connected. In References [6,32], the lifetime definitions rely on the percentage of covered area in the monitored site. But they are not suitable for environmental applications which are data-driven. In data-driven applications, we are more interested in sampling data from the monitored site rather than providing full coverage. Tightly connecting these data gathering nodes can reduce redundant information transmissions, and, hence prolong the overall network lifetime.

Motivated by the benefits of device heterogeneity, as well as the 3-D grid model, our research provides an efficient grid-based deployment for provisioning WSNs of maximum backbone (critical nodes) connectivity degree under lifetime constraints and limited cost budget in environmental applications. For more efficiency and unlike other grid-based deployments, we find the most feasible grid vertices to be searched for are those with optimal deployment rather than searching a massive number of grid vertices in large-scale environmental applications.

3.2 System Models and Problem Definition

In this section, we outline our assumed wireless sensor network models. Additionally, we introduce a general definition for the targeted relay node deployment problem. We assume hierarchical network architecture to address the node heterogeneity problem. A graph topology is considered for easy network extension and accurate (mathematical) connectivity computation. Furthermore, a detailed discussion of the utilized cost and communication models is proposed, and the appropriateness of the considered lifetime definition is examined.

3.2.1 Network Model and Placement Problem

In this chapter, a two-layer hierarchical architecture is assumed to be a natural choice in large-scale environmental applications as well as providing a more energy-efficient deployment plan. The lower layer consists of sensor nodes that sense the targeted phenomena and send measured Rdata to cluster heads (CHs) in the upper layer, as shown in Figure 3.1a. These sensor nodes usually have fixed and limited transmission ranges and do not relay traffic in order to conserve more energy. The upper layer consists of cluster heads and relay nodes which have better transmission range ($=r$) and communicate periodically with the base station to deliver the measured data in the lower layer. Cluster heads aggregate the sensed data and coordinate the medium access; they also support relay nodes in relaying data from other CHs to the BS in the upper layer. Assuming sensor nodes have enough energy to perform their effortless tasks, we focus this work on the upper layer devices which are relay nodes and cluster heads. The topology of the upper layer is modeled as a graph $G=(V, E)$, where $V=\{n_0, n_1, ..., n_{nc}\}$ is the set of n_c candidate grid vertices, E is the set of edges in graph G, and $(i, j) \in E$, if nodes at n_i and n_j have enough probabilistic connectivity percentage to establish a communication

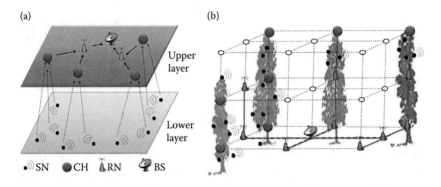

(a)   (b)

Upper layer

Lower layer

•)) SN   ● CH   ⚡ RN   🛰 BS

Figure 3.1   (a) Two-layer hierarchical architecture and (b) Cubic 3-D grid model for the targeted wireless sensor network deployment where dashed lines and empty circles represent grid edges and vertices, respectively.

link (edge). We assert that deployment of relay nodes in this research is independent of the underlying Medium Access Control (MAC) protocol, where we assume a transmission rate limit T for each node during a one-time unit (measured in days). This limit can be adjusted to comply with any MAC protocol. For simplicity and without losing generality, we assume S-MAC protocol [35] is handling the medium access in this research with a 50% duty cycle and with only 6 bytes controlling fields in the exchanged packets for more energy savings. Moreover, the assumed traffic generation model is flexible enough to support different requirements of numerous WSN applications as we have specific parameters, T and Y, controlling the assumed packet transmission rate and arrival rate respectively. For example, if we increase T, packets will be generated more frequently. If we decrease Y, packets will take more time to reach the destination; thus, a more congested network can be experienced. In this research, we assume that the arrival rate Y is following a passion process which is very natural in WSN simulations.

Figure 3.1b depicts the 3-D grid model assumed in this chapter; the grid edge length is supposed to be equal to a relay node transmission range r. It is assumed that all relay nodes have a common transmission range r. We think our deployment planning is applicable for other types of grid models, not just the cubic one. In this cubic grid model, each sensor node (SN) is placed near to phenomena of interest for more accurate estimates in terms of the spatial properties of the collected data. Cluster heads (CHs) are then placed on the most appropriate grid vertices, which can serve the largest number of sensor nodes distributed around each cluster head. The base station is placed based on the application requirements in a fixed position and is the data sink for the system. Then, we seek to optimize the relay node positions on the 3-D grid to get cluster heads connected to the base station efficiently, that is, in terms of cost and network lifetime. Hence, we define a general relay node placement problem in environmental wireless sensor networks as follows:

Problem Statement Definition: *Given a specific sensing task with pre-specified SNs, CH and BS locations, determine the positions of RNs so that connectivity between CHs and BS is maximized while lifetime and cost constraints are satisfied.*

3.2.2 Cost and Communication Models

Device cost in environmental applications depends on its functionalities and hardware components. The more functionality the device has, the more complex and expensive it is. As relay nodes are assumed to have more functionality and dominate other devices in terms of transmission range, the cost is modeled in this chapter by the number of relay nodes placed in the monitored site. We assume identical cost for these RNs.

For the communication model, we consider a probabilistic connectivity between the deployed devices, in which wireless signals not only decay with distance, but also are attenuated and reflected by surrounding obstacles such as trees, animals, hills, etc. Accordingly, the communication range of each device must be represented by an arbitrary shape as depicted in Figure 3.2. For realistic estimation of the ability to communicate within this arbitrary shape, we need a signal propagation model that reflects the effects of the surrounding obstacles and the environmental characteristics on the propagated signals. This model can describe the path loss* in the monitored environment as follows [28].

$$P_r = K_0 - 10\gamma \log(d) - \mu d \tag{3.1}$$

where P_r is the received signal power, d is the Euclidian distance between the transmitter and receiver, γ is the path loss exponent calculated based on experimental data, μ is a random variable that

Figure 3.2   An arbitrary shape of the communication range in 3-D space, due to attenuation and shadowing affecting outdoor wireless signals.

---

* Path loss is the difference between the transmitted and received power of the signal.

follows a log-normal distribution function with zero mean and variance δ^2 to describe signal attenuation effects* in the monitored site, and K_0 is a constant calculation based on the transmitter, receiver, and monitored site mean heights.

Let P_r equal the minimal acceptable signal level to maintain connectivity. Assume γ and K_0 in Equation 3.1 are also known for the specific site to be monitored. Thus, a probabilistic communication model which gives the probability that two devices separated by distance d can communicate with each other is given by

$$P_c = Ke^{-\mu d^\gamma} \tag{3.2}$$

where $K_0 = 10\log(K)$

Thus, the probabilistic connectivity P_c is not only a function of the distance separating the wireless nodes but also a function of the surrounding obstacles and terrain, which can cause shadowing and multipath effects (represented by μ). Thus, the ability to communicate between two nodes is defined as follows:

Probabilistic Connectivity Definition: *Two nodes (devices) i and j, separated by distance d, are connected with a threshold parameter τ ($0 \leq \tau \leq 1$), if $P_c(i, j) \geq \tau$.*

Note that this communication model is generic in terms of the parameters (K, μ, τ, and γ), which specify the surrounding environment characteristics. Setting these parameters to values obtained from experimental data would provide more practical connectivity estimation, and thus more efficient deployment planning.

3.2.3 Lifetime Model

Models in the literature differ in the way they consider a wireless sensor network to be still operational. These models can rely on connectivity of the deployed nodes or on percentage of alive nodes (which have enough energy to accomplish their assigned tasks) in the network. *Connectivity-based (CB)* and *Percentage of Alive Nodes (PAN)* models are defined as follows:

---

* Wireless signals are weakened because of shadowing and multipath effects. This refers to the fluctuation of the average received power.

Connectivity-Based Definition: *Lifetime of a WSN is the time span from deployment to the instant when a network partition occurs*.*

Percentage of Alive Nodes Definition: *Lifetime of a WSN is the time span from deployment to the instant when the percentage of alive nodes falls below a specific threshold.*

However, losing a few nodes may not significantly affect the overall wireless sensor network performance especially when redundant nodes and communication links (edges) are used in tolerating high probabilities of disconnected nodes as in environmental applications. In environment monitoring, several nodes are generally assigned to measure single specific criteria of the monitored space, such as temperature in forestry fire detection [31]. Consequently, the concept of node redundancy should be addressed. Furthermore, lifetime models relying on the above definitions do not take into consideration the node type which could be cluster head, relay node, or sensor node. Therefore, we propose the following *Environment-specific (Env.)* lifetime definition.

Environment-Specific Lifetime Definition: *Lifetime of a WSN is the time span from deployment to the instant when the percentage of alive and connected irredundant nodes is below a pre-defined specific threshold.*

Using this definition, we benefit from device redundancy by considering the network to be operational as long as a specific percentage of CHs providing the targeted data is still alive. These CHs need not only be alive but also must be connected to the base station via single or multi-hop path(s). Note that a cluster head i is connected to another node j, if $P_c(i, j) \geq \tau$, according to the *probabilistic connectivity* definition.

In order to mathematically translate the aforementioned lifetime definitions, we assume *number of rounds* for which a wireless sensor network can stay operational as the unit measure of the network lifetime. A complete *round* is defined in this chapter as the time span t_{round} in which each irredundant cluster head (i.e., responsible for different sensor nodes) transmits at least once to the base station without violating cutoff criteria of the lifetime definitions. t_{round} is identical for all rounds due to a constant data delivery assumed per round. In addition,

---

* Network partition occurs when one or more nodes are not able to communicate with the base station.

we adopt the general energy consumption model proposed in Reference [35] in which energy consumed for receiving a packet of length L is

$$J_{rx} = L\beta \qquad (3.3)$$

and the energy consumed for transmitting a packet of length L for distance d is:

$$J_{tx} = L\left(\varepsilon_1 + \varepsilon_2 d^\gamma\right) \qquad (3.4)$$

where ε_1, ε_2 and β are hardware specific parameters of the utilized transceivers, and γ is the path loss exponent.

Based on Equations 3.3 and 3.4, in addition to knowing the initial energy E_i of each node with its relative position to other nodes, we can calculate the remaining energy E_r per node after the completion of each round by

$$E_r = E_i - TJ_{tx} - RJ_{rx} - AJ_a \qquad (3.5)$$

where T, R, and A are the arrival rates of transmitted, received and aggregated packets per round respectively. This follows a Poisson distribution, and J_a is the energy consumed for a single packet aggregation. Considering E_r calculated in Equation 3.5 and assuming the cutoff criterion associated with each lifetime definition is represented by a binary variable[*] C, we can calculate the total number of rounds for which a wireless sensor network can stay operational.

To assess the *environment-specific* definition, we use simulation to compare it to *connectivity-based* and *percentage of alive nodes* definitions using four main performance metrics: *(1) percentage of alive CHs, (2) percentage of disconnected CHs/RNs, (3) ratio of remaining energy (RRE), and (4) total rounds, Percentage of alive CHs* is the percentage of cluster heads which have enough energy to aggregate and forward data to the base station at least once. *Percentage of disconnected CHs/RNs* is the percentage of cluster heads and relay nodes which have enough energy to aggregate and forward data at least once but are not able to communicate with the base station. *Ratio of remaining energy* is the ratio of total energy amount still available at all nodes (CHs/RNs)

---

[*] The cutoff criterion is not satisfied and the network is still considered operational if $C = 0$ and vice versa if $C = 1$.

to the total energy at deployment when the network is not operational. The network is not operational when the cutoff criterion of the lifetime definition is satisfied. Finally, *total rounds* is the total number of rounds in which a wireless sensor network can be considered operational. These four performance metrics are chosen to reflect the ability of *environment-specific* lifetime definition to (1) accurately estimate the network lifetime and (2) effectively utilize the network (i.e., maximize network operational time by delaying the assumption of the network death.) Thus, better energy and resource utilization can be achieved. Using MATLAB®, we simulate randomly generated wireless sensor networks which have the hierarchical architecture and the graph topology proposed in Section 3.3.1. Each generated network consists of 12 CHs and a total of 50–80 RNs which are randomly deployed on grid vertices in $700 \times 700 \times 200$ (m³) 3-D space using a Linear Congruential random number generator. The parameters used in the simulations are listed in Table 3.1.

Therein, τ is set to high value for practicality in simulating fluctuated and attenuated signals in environment monitoring applications [28]. For simplicity, we apply cubic 3-D grid model with identical grid edges of the length equal to 100 (m). We assume a predefined fixed time schedule for traffic generation (=100 packets per round from each CH) and a (PNF) varying from 10% to 60%. We define the PNF as the probability of physical damage for each node in the network which is very common in outdoor environment monitoring. Thus, a higher PNF indicates a higher possibility for the node to be damaged while still having enough energy to sense and communicate. We intend to assume a very high PNF (up to 60%) to reflect

Table 3.1   Parameters of the Simulated WSNs

PARAMETER	VALUE	PARAMETER	VALUE
T	70%	L	512 (bits)
n_c	110 (vertex)	E_i	15.4 (J)
ε_1	50e−9 (J/bit)	T	100 (packet/round)
ε_2	10e−12 (J/bit/m²)	P_r	−104 (dB)
β	50e−9 (J/bit)	t_{round}	24 (h)
γ	4.8	K_0	42.152
δ^2	10	r	100 (m)
J_a	50e−7 (J)	PNF	10%–60%

some actual situations in outdoor environmental applications. Three different cutoff criteria are used for the simulated network to be considered operational. According to *environmental-specific* (*Env.*) lifetime definition and the proposed network model, the network is still operational as long as the percentage of connected irredundant cluster heads, which have enough energy to communicate with the base station, is greater than or equal to 50%. The *CB* definition considers the network nonoperational when one or more irredundant cluster heads are unable to reach the base station. Finally, the network is not operational, based on the *PAN* definition, when 50% or more of the nodes run out of energy. After the network is considered not operational, we measure these performance metrics. This experiment is repeated 500 times. The average results are reported in Figures 3.3 through 3.6. We observe that the average results hold a confidence interval no more than 5% of the average (over 500 runs) at a 95% confidence level.

Figure 3.3 shows the percentage of disconnected nodes obtained when the network becomes nonoperational. Obviously, *CB* definition underestimates the network lifetime by considering it nonoperational

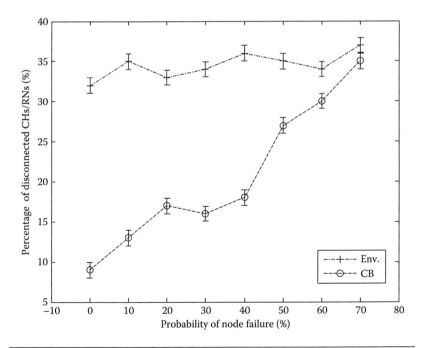

Figure 3.3 Percentage of disconnected CHs/RNs vs. different probabilities of CH/RN failure.

while it has very low percentage of disconnected nodes. Dependency of the disconnected nodes percentage on the PNF values confirms its unsuitability for environmental applications. Using *Env.* definition we observe that the simulated wireless sensor networks can remain operational even when the percentage of disconnected nodes is much higher than the percentage achieved by the *CB* definition. Unlike the *CB* definition, the *Env.* definition shows very close percentages of disconnected nodes under different probabilities of node failure when the network is considered nonoperational. This steady state in lifetime estimation is preferred under the varying probabilities of node failures in outdoor environments. In Figure 3.4, the percentage of alive CHs when the network becomes nonoperational is shown. In this figure, we can see how definitions relying on percentage of alive nodes waste network resources by stopping the network while it still has significant percentage of alive CHs. Furthermore, the *PAN* definition considers that the network is not operational based on an unpractical percentage of alive nodes which ignore node type (whether it is a CH

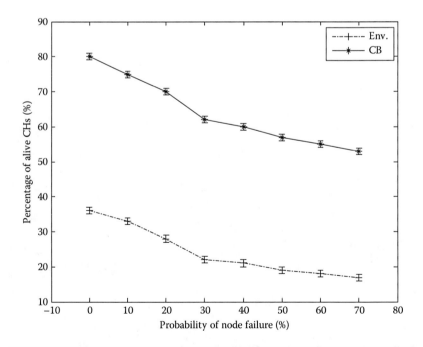

Figure 3.4   Percentage of alive CHs vs. different probabilities of CHs/RNs failure.

or an RN) and does not differentiate between redundant and irredundant CHs. Therefore, *PAN* is not suitable for environmental applications. Figure 3.5 shows that based on the *Env.* definition the wireless sensor network can remain operational even when the RRE is less than 20%. On the other hand, using the *CB* and *PAN* definitions, the network is nonoperational even when it has around 60% of the initial deployed energy. Thus, Figures 3.4 and 3.5 indicate how irrelevant lifetime definitions may lead to severe waste in resources (e.g., functional nodes, remaining unexploited energy, etc.) by assuming the network is dead while it still has the ability to continue its assigned task. Figure 3.6 depicts the total rounds counted based on the three lifetime definitions. It elaborates on how lifetime models relying on the *CB* and *PAN* definitions can underestimate the overall network lifetime. It indicates the appropriateness of the *Env.* definition in practice under harsh outdoor operational conditions. Consequently, the *Env.* lifetime definition is imposed in our deployment strategy and is described in the following section.

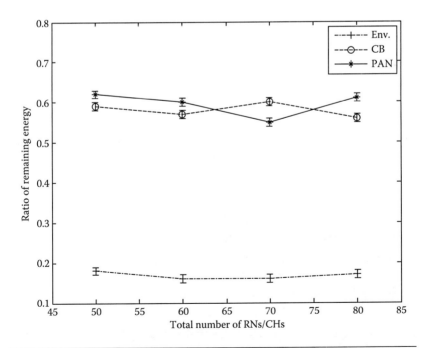

Figure 3.5 Comparison of the three different definitions in terms of Ratio of Remaining Energy (RRE).

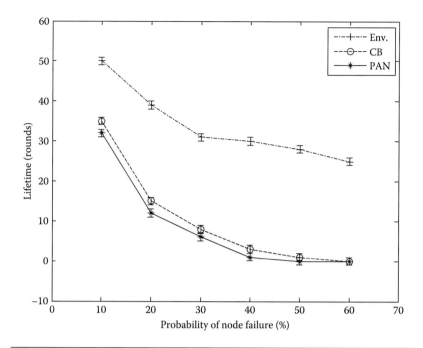

Figure 3.6 Comparison of the three different definitions in terms of total rounds.

3.3 Deployment Strategy

The relay node placement problem proposed in this chapter has an infinitely large search space and finding the optimal solution is highly nontrivial. Therefore, we propose a 3-D grid model that limits the search space to a more manageable size. Grid models have well organized vertices distributed in regular lattice structures. These vertices can be organized in different structures (e.g., cubes, octahedrons, pyramids, etc.) in a 3-D space to provide more accurate estimates in terms of the spatial properties of the targeted data.

We assume knowledge of the 3-D terrain of the monitored site ahead of the deployment planning time. Hence, practical candidate positions on the grid vertices are predetermined; unfeasible positions are excluded from the search space. We use candidate grid vertices to apply the deployment strategy in two phases. The first phase is used to place a minimum number of relay nodes on the grid vertices to establish a connected network. The second phase is used to choose the optimal positions of extra relay nodes required to maximize the network connectivity with constraints on cost and lifetime. This two-phase deployment strategy is called *Optimized 3-D Grid Deployment with Lifetime Constraint (O3DwLC).*

3.3.1 First Phase of the O3DwLC Strategy

The first phase is achieved by constructing a connected backbone (B) using first phase relay nodes (FPRNs). Locations of these nodes are optimized in order to use a minimum number of FPRNs that can connect CHs to the BS. Toward this end, we apply the Minimum Spanning Tree (MST) algorithm as described in Algorithm 3.1.

Algorithm 3.1: MST to construct the connected backbone (B)

1. **Function ConstructB** (IS: Initial Set of nodes to construct B)
2. **Input:**
3. A set *IS* of the CHs and BS nodes' coordinates
4. **Output:**
5. A set *CC* of the CHs, minimum RNs, and BS coordinates forming the network Backbone
6. **Begin**
7. CC=set of closest two nodes in IS;
8. $CC = CC \cup$ minimum RNs needed to connect them on the 3-D grid;
9. $IS = IS - CC$;
10. N_d=number of remaining *IS* nodes which are not in *CC*;
11. $i=0$;
12. **for each** remaining node n_i in *IS* **do**
13. Calculate M_i: Coordinates of minimum number of RNs required to connect n_i with the closest node in CC^*.
14. $i=i+1$;
15. **end**
16. $M=\{M_i\}$

---

* This is achieved by counting the minimum number of adjacent grid vertices, which establish a path from the separated CH at vertex n_i to a currently Connected Component CC.

17.   **while** $N_d > 0$ **do**

18. $SM = $ Smallest M_i;

19. $CC = CC \cup SM \cup n_i$;

20. $IS = IS - n_i$;

21. $M = M - M_i$;

22. $N_d = N_d - 1$;

23.   **end**

24.   **end**

Algorithm 3.1 aims at constructing the MST using the grid vertices representing the 3-D space candidate positions. Line 7 of Algorithm 3.1 search for the closest[*] two nodes in the initial set IS, which have the CHs and BS. If the closest two nodes are not adjacent on the 3-D grid (i.e., $P_c \leq \tau$), line 8 adds the minumum number of grid vertices on which the relays have to be placed to establish a path between these two nodes. After connecting the closest two nodes (i.e., establishing a connected component CC), we iterativelly look for the next closest node that has to be connected to the CC. This has been achieved through lines 12–22 of Algorithm 3.1.

For more elaboration on placement of the FPRNs using Algorithm 3.1, consider the following example.

> **Example 3.1:** Assume we have 7 cluster heads preallocated with the BS on the grid vertices as in Figure 3.7a. We then seek the minimum number of RNs ($=N_{MST}$) required to connect these CHs with the BS as depicted in Figure 3.7b. Positions of these N_{MST} relay nodes are determined by applying Algorithm 3.1. The algorithm first constructs the initial set IS consisting of cluster heads and base station coordinates. Then, a connected component CC set is initiated by the closest two node coordinates in the IS; in this example, these are the cluster heads at vertices 15 and 17. These coordinates are then removed from the IS. Obviously, by adding only one relay node at vertex 14, cluster heads at 15 and 17 become connected. Hence, coordinates of that relay node is added to the CC and the remaining number of nodes N_d in IS set to 6. Now, we calculate M_i for the remaining nodes at vertices 1, 5, 19,

---

[*] In terms of the vertices count separating the two nodes.

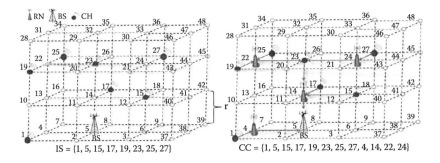

Figure 3.7   An example of FPRNs placement using Algorithm 3.1. (a) Before applying Algorithm 3.1 (i.e., without FPRNs) and *r* is the grid edge length and (b) after applying Algorithm 3.1 (i.e., with FPRNs).

23, 25 and 27 in *IS*, which would have 2, 0, 2, 0, 1 and 1 candidate relay node coordinates respectively. Since the set M_1, associated with the base station placed at vertex 5, has the smallest number of required coordinates (=0), we put M_1 in the set *SM* and *CC* becomes equal to {15, 17, 14, 5}. *M* and *IS* are then updated and N_d is decremented by 1. By repeating this process until N_d is equal to 0, we obtain the final connected component *CC* that is shown in Figure 3.7b, where relay nodes in this figure are the FPRNs of the wireless sensor network to be deployed. The deployed FPRNs with the cluster heads and the base station construct the network backbone.   ∎

Connectivity of the backbone *B* generated in this phase of the deployment is measured by considering *B* as a connected graph which has a Laplacian matrix *L(B)* [14]. The Laplacian matrix is a 2-D matrix that has −1 at the element (i,j) if there is a connection between nodes *i* and *j*. It has an integer positive number at the element (i,i) that represents the number of edges connected to the node *i* (see Figure 3.8). Given *L(B)*, the backbone connectivity (or algebraic connectivity) is mathematically measured by computing the second smallest eigenvalue λ_2. Where λ_2 indicates the minimum number of nodes and links whose removal would disconnect the graph *B* (see Figure 3.8 for more elaboration on λ_2). By maximizing λ_2 of *L(B)*, we maximize the required number of nodes and communication links to disjoint (disconnect) paths in the network backbone. This is because of the proportional relationship between the value of λ_2 and the number of nodes/links which can cause network partitions according to Figure 3.8. Hence,

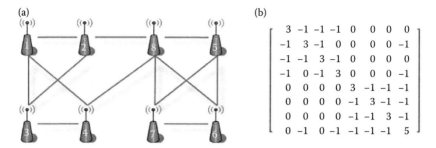

Figure 3.8   (a) A graph with 8 nodes and 13 links. The graph's connectivity characteristics are: one node to disconnect (removal of node 8), two links to disconnect (removal of links connecting node 8 to nodes 2 and 4), Laplacian matrix of this graph is shown in (b) and λ_2 of the matrix in (b) is equal to 0.6277. As λ_2 increases the node/link count required to partition the network increases.

more reliable[*] environmental wireless sensor networks can be achieved due to the ability to overcome significant topology changes caused by communication quality changes and node failures using tightly connected backbones. In order to *maximize* the backbone connectivity λ_2, extra relay nodes (SPRNs) are placed in the second phase of the O3DwLC strategy.

3.3.2 Second Phase of the O3DwLC Strategy

In this phase, we optimize positions of SPRNs such that λ_2 of the backbone generated in the first phase is maximized with constraints on cost and lifetime. For simplicity, we start by maximizing λ_2 without lifetime constraints. Assume we have n_c grid vertices as candidate positions for SPRNs. We want to choose the optimum N_{SPRN} relay nodes amongst these n_c relays with respect to connectivity, where N_{SPRN} is constrained by a cost budget. We can then formulate this optimization problem, with reference to Table 3.2, as

$$\max \lambda_2 \left(L(\alpha) \right)$$

$$s.t. \quad \sum_{i=1}^{n_c} \alpha_i = N_{SPRN}, \; \alpha_i \in \{0, 1\}, \tag{3.6}$$

---

[*] Reliability here is defined by the existence of an operational path from all CHs to the BS even in the presence of nodes and link failures.

Table 3.2 Notations Used in the Placement Problem

NOTATION	DESCRIPTION
α_i	A binary variable equals 1 when RN at vertex i in the 3-D grid is allocated and 0 otherwise.
A_i	Incidence matrix that results by adding RN_i in the 3-D grid; $A_i = [a_1, a_2, \ldots, a_m]$, where a_i is the vector that consists of n elements that can take a value of either 0, 1 or −1 and m is the total number of edges that is produced by adding RN_i. For example, if adding RN_i will establish a connection between node 1 and 3, then the 1st element is set to 1 and the 3rd element is set to −1 and all remaining elements are set to zeros.
N	Summation of $N_{CH} + N_{MST} + 1$ (which is the total of CHs, FPRNs, and BS).
L_i	Initial Laplacian matrix produced by the allocated CHs, FPRNs, and the BS nodes.
$I_{n \times n}$	Identity matrix of size n by n.

where

$$L(\alpha) = L_i + \sum_{i=1}^{n_c} \alpha_i A_i A_i^T \qquad (3.7)$$

However, an exhaustive search scheme is required to solve 3.6, which is computationally expensive, especially for the naturally large n_c values in large-scale environmental applications. This is due to the involved computations required for finding λ_2 for a large number

$$\left(= \binom{nc}{N_{SPRN}} \right)$$ of Laplacian matrices. Therefore, we need a compu-

tationally efficient means to solve 3.6 as well as a more limited search space that reduces the value of n_c.

Taking advantage of the constructed network backbone in the first phase, SPRNs may be placed on any grid vertex as long as it is within the probabilistic communication range of the largest number of CHs/FPRNs in B. This, in turn, can further finite the search space without affecting the quality of the deployment plan. To explain our method of finding such a finite search space, we introduce the following definitions.

Ideal Set Definition: *A finite set of positions P is ideal if it satisfies the following property:*

There exists an optimal* placement of SPRNs in which each relay is placed at a position in P.

---

* Optimal in terms of connectivity.

We aim at finding such an ideal set in order to achieve more efficient discrete search space in which candidate relay positions are not including all of the grid vertices, but a subset of these vertices that has the most potential to enhance the network connectivity. Moreover, since the computational complexity will be proportional to the cardinality of this ideal set, we should find a set with reasonably small size.

Covered Grid Unit (CGU) Definition: *A covered grid unit α is a grid unit that has a connected center with at least one CH or FPRN. Let C(α) denote the subset of the CHs/FPRNs coordinates covering α.*

We assume that the considered virtual 3-D grid can have a building unit, called *grid unit.* For example, the *grid unit* of the 2-D grid shown in Figure 3.9 is the small square drawn by dashed lines. while in 3-D it will be a cube (in cubic grid models.) Each *grid unit* is supposed to have a center of mass represented by its position coordinates (black dots in Figure 3.9.) We determine that a *grid unit* is covered by a specific node (CH/FPRN) when the probabilistic connectivity P_c between the *grid unit* center and that node is greater than or equal to the aforementioned threshold τ.

Maximal Covered Grid Unit (MCGU) Definition: *A covered grid unit α is maximal if there is no Covered Grid Unit β, where $C(\alpha) \subseteq C(\beta)$.*

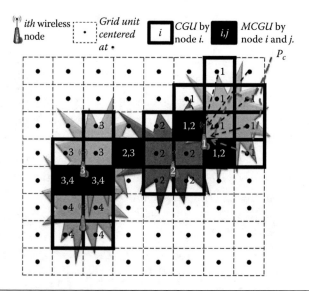

Figure 3.9 An example of a maximum covered grid unit in 2-D plane. Numbers inside the bounded squares represent the node ID covering these squares.

For more illustration, consider Figure 3.9 in the 2-D plane. It shows four wireless nodes with respect to the grid units. Each wireless node has an arbitrary communication range. The covered grid units are bounded by solid lines, and the *Maximal Covered Grid Units (MCGUs)* are solid black squares. These MCGUs have the highest potential to place the SPRNs due to their ability to establish the highest number of new edges between already deployed FPRNs/CHs. Accordingly, we have to show that an ideal set can be derived from the set of MCGUs. Toward this end, we state the following Lemmas:

Lemma 3.1: For every CGU β, there exists an MCGU α such that $C(\beta) \subseteq C(\alpha)$.

Proof: If β is an MCGU, we choose α to be β itself. If β is not an MCGU then, by definition, there exists a covered grid unit α_1 such that $C(\beta) \subseteq C(\alpha_1)$. If α_1 is an MCGU, we choose β to be α_1, and if α_1 is not maximal then, by definition, there exists another covered grid unit α_2 such that $C(\alpha_1) \subseteq C(\alpha_2)$. This process continues until a maximal covered grid unit α_x is found; we choose α to be α_x. Thus, Lemma 3.1 holds. ∎

Lemma 3.2: Finding an MCGU takes at most $(n-1)$ step, where n is number of nodes constructing the backbone B.

Proof: By referring to the proof of Lemma 3.1, it is obvious that $|C(\alpha_x)| \le n$, and $|C(\alpha)| < |C(\alpha_1)| < |C(\alpha_2)| < \cdots < |C(\alpha_x)| \le n$; where $|C|$ represents the cardinality of the set C. Consequently, the process of finding the maximal covered grid unit α_x takes a finite number of steps less than or equal to $n-1$. ∎

Then, we introduce the following Theorem:

Theorem 3.1: A set P that contains one position from every MCGU is ideal.

Proof: To prove this Theorem, it is sufficient to show that for any arbitrary placement Z we can construct an equivalent[*] placement

---

[*] Equivalent in terms of the covering CHs/FPRNs. In other words, the placement of a SPRN at position i, within the communication range of the CHs x and y, is equivalent to the placement of the same SPRN node at position j within the communication range of the CHs x, y and z.

Z' in which every SPRN is placed at a position in P. To do so, assume that in Z, a SPRN i is placed such that it is connected to a subset J of CHs/FPRNs. It is obvious that there exists a covered grid unit β, such that $J \subseteq C(\beta)$. From *Lemma 3.1,* there exist an MCGU α such that $C(\beta) \subseteq C(\alpha)$. In Z', we place i at the position in P that belongs to α, so that i is placed at a position in P and is still connected with all CHs/FPRNs in J. By repeating for all SPRNs, we construct a placement Z' which is equivalent to Z, and thus Theorem 3.1 holds.   ■

In order to find all MCGUs, we need a data structure associated with each *grid unit* to store coordinates and total number of CHs/FPRNs covering the grid unit. We represent this data structure by the *covered grid unit* set $C(i)$, where i is the center of the grid unit. By computing $C(i)$, $\forall\ i \in V$, we can test whether a covered grid unit centered at i is maximal or not by searching for a set that has at least all elements of $C(i)$. In the following, Algorithm 3.2 establishes the data structures in $O(n)$. It associates each grid unit center with its total covering nodes. In line 9 of Algorithm 3.2, we compute the probability of the grid unit center i being connected with each backbone node individually based on Equation 3.2. This is repeated by lines 5–15 until all probabilities between the grid unit centers and all backbone nodes are computed and the data structure is formed. Algorithm 3.3 tests whether a *covered grid unit* is maximal or not. In line 11 of Algorithm 3.3, we search for any set (other than C_i) in C that has the same backbone nodes which cover the grid center i. If such a set is found, Algorithm 3.3 returns false; otherwise, it returns true, meaning that set C is maximal. Algorithm 3.4 uses Algorithms 3.2 and 3.3 to construct an ideal set P by finding all MCGUs. The overall complexity of Algorithm 3.4 is $O(n \log n)$.

Algorithm 3.2: Build up grid unit data structure

---

1. **Function FindGridUnitCoverage** (B: Backbone constructed by CHs and FPRNs)
2. **Input:**
3. A set B of the CHs and FPRNs nodes' coordinates.
4. **Begin**

5. **foreach** grid unit center i **do**
6. $C(i) := \varnothing$;
7. **foreach** CH/FPRN j **do**
8. Compute $P_c(i, j)$;
9. **If** $P_c(i, j) \geq \tau$
10. $C(i) := j \cup C(i)$;
11. **endif**
12. **endfor**
13. **endfor**
14. **End**

Algorithm 3.3: Testing whether a grid unit set $C(i)$ is maximal or not

---

1. **Function Maximal**($C(i)$, all nonempty grid unit sets)
2. **Input:**
3. A set $C(i)$ for a specific grid unit center i.
4. All nonempty sets of the grid unit centers.
5. **Output:**
6. True if $C(i)$ is MCGU and False otherwise.
7. **Begin**
8. **If** $C(i) := \varnothing$ **do**
9. **return** False;
10. **endif**
11. Search for a set C' such that $C(i) \subseteq C'$.
12. **If** $C' := \varnothing$ **do**
13. **return** True;
14. **else**
15. **return** False;
16. **endif**
17. **End**

Algorithm 3.4: Finding all *Maximal Covered Grid Units* MCGUs

---

1. **Function FindMCGUs** (*B*: Backbone constructed by CHs and FPRNs)
2. **Input:**
3. A set *B* of the CHs and FPRNs coordinates.
4. **Output:**
5. A set *P* that contains one position from every MCGU.
6. **Begin**
7. $P := \emptyset$;
8. **FindGridUnitCoverage** (*B*);
9. **foreach** *C*(*i*) **do**
10. **If Maximal**(*C*(*i*), all nonempty grid unit sets) **do**
11. $P := \{i\} \cup P$;
12. **endif**
13. **endfor**
14. **End**

Once we obtain the set *P* which contains one position (grid vertex coordinates) from each MCGU, the search space of the problem formulated in 3.6 becomes much more limited.

In order to efficiently solve the optimization problem in 3.6, we reformulate it as a standard semi-definite program (SDP) optimization problem [7,14], which can be solved using any standard SDP solver. By relaxing the Boolean constraint $\alpha \in \{0, 1\}$ to be linear constraint $\alpha \in [0, 1]$, we can represent the problem in 3.6 as

$$\max \lambda_2 \left(L(\alpha) \right)$$

$$s.t. \quad \sum_{i=1}^{n_c} \alpha_i = N_{SPRN}, \quad 0 \le \alpha_i \le 1, \tag{3.8}$$

The optimization problem in 3.8 is convex with linear constraint [7]. Therefore, we introduce the following Theorem:

Theorem 3.2: The optimization problem in 3.8 is mathematically equivalent to the following SDP optimization problem.

max S

$$s.t. \quad S\left(I_{nxn} - \frac{1}{n}11^T\right) \preceq L(\alpha), \quad \sum_{i=1}^{n_c} \alpha_i = N_{SPRN}, \quad 0 \le \alpha_i \le 1, \qquad (3.9)$$

where S is a scalar variable and $\preceq$ denotes the positive semi-definiteness (i.e., all eigenvalues of the matrix are greater than or equal to zero).

Proof: Let $V \in \mathbf{R}^n$ be the corresponding eigenvector of $\lambda_2(L(\alpha))$. Thus, $1^T V = 0$, and $V = 1$.

Since,

$$L(\alpha)V = \lambda_2 V \qquad (3.10)$$

Hence,

$$V^T L(\alpha)V = \lambda_2 V^T V = \lambda_2 \qquad (3.11)$$

$$\Rightarrow \lambda_2(L(\alpha)) = \inf_V \{V^T L(\alpha)V \mid 1^T V = 0, \text{ and } V = 1\}^* \qquad (3.12)$$

Let

$$L'(\alpha) = L(\alpha) - S\left(I_{nxn} - \frac{1}{n}11^T\right) \qquad (3.13)$$

Thus for any $V_{n\times 1}$ where $1^T V = 0$, and $V = 1$, we get

$$V^T L'(\alpha)V = V^T L(\alpha)V - S\left(V^T I_{nxn}V - \frac{1}{n}(V^T 1)(1^T V)\right)$$

$$= V^T L(\alpha)V - S \qquad (3.14)$$

Hence, for $L'(\alpha)$ to be positive semi-definite, the maximum value of S should be

$$S = \inf_V \{V^T L(\alpha)V \mid 1^T V = 0, \text{ and } V = 1\}, \qquad (3.15)$$

From Equations 3.12 and 3.15,

$$S = \lambda_2(L(\alpha)) \qquad (3.16)$$

---

* $\lambda_2(L(\alpha))$ is the point-wise infimum (i.e., lower bound) of a family of linear functions of α. Hence, it is a concave function in α. In addition, the relaxed constraints are linear in α. Therefore, the optimization problem in 3.8 is convex.

Therefore, maximizing S in Equation 3.9 is equivalent to maximizing $\lambda_2\left(L(\alpha)\right)$ in 3.8 if the constraints are satisfied. ∎

In order to add environmental lifetime constraints to Equation 3.9, let the backbone (generated in the first phase) be operational for the initial number of rounds equal to IRs. Assume adding one relay node of the SPRNs would prolong the network lifetime by extra rounds ER_i. To guarantee that the network will stay operational for a minimum number of required rounds RLT, the total extra and initial rounds must be greater than or equal to RLT as elaborated in the following:

$$-\sum_{i=1}^{n_c} ER_i \leq (IRs - RLT) \tag{3.17}$$

Since we are using the cutoff criterion of the *environmental lifetime* definition in calculating both ER_i and IRs, inequality 3.17 represents a more environment-specific lifetime constraint in the O3DwLC strategy.

From 3.9 and 3.17, SPRNs positions that maximize λ_2 with constraints on lifetime and cost are found by solving the following[*]:

$$\max S$$

$$s.t. \quad S\left(I_{nxn} - \frac{1}{n}11^T\right) \leq L(\alpha), \sum_{i=1}^{n_c}\alpha_i \leq N_{SPRN},$$

$$-\sum_{i=1}^{n_c} ER_i\alpha_i \leq (IRs - RLT), \ 0 \leq \alpha_i \leq 1, \tag{3.18}$$

In the following, Algorithm 3.5 summarizes the second phase deployment proposed in this section where the search space is limited to n_c positions for grid vertices within the ideal set P.

Algorithm 3.5: SPRNs deployment

1. **Function SPRNs** (*B*: Backbone constructed by CHs, FPRNs and BS, *P*)
2. **Input:**
3. A set B of the CHs, FPRNs and BS nodes' coordinates

---

[*] SDPA-M MATLAB package can be used to solve (3.18).

4. An ideal set P of n_c candidate positions for the SPRNs

5. **Output:**

6. A set SP of the SPRNs coordinates maximizing connectivity of B with practical lifetime and cost constraints

7. **Begin**

8. L_i=Laplacian matrix of B

9. **IRs**=number of rounds B can stay operational for

10. **for** (i=1; i<n_c; i++)

11. A_i=coefficient matrix corresponding to vertex i on the grid

12. ER_i=extra rounds achieved by allocating RN at vertex i

13. **end**

14. **SP**=Solution of SDP in 3.18

15. **End**

For more elaboration on Algorithm 3.5, consider the following example:

Example 3.2: Assume we have up to two extra relay nodes (SPRNs) to maximize connectivity of the backbone generated in Figure 3.7b to ensure the network can stay operational for at least 20 rounds. In this case, $N_{SPRN} = 2$, $n = 12$, and $RLT = 20$. We start by computing the ideal set P to specify our search space in this problem using Algorithm 3.4. Afterward, we calculate the initial Laplacian matrix L_i associated with the backbone to be used in Equation 3.7 in addition to computing initial rounds (IRs=10) for which the backbone can stay operational. With reference to Table 3.3, we set α_i to 1 and calculate A_i and extra rounds ER_i for each element i in the ideal set P. Notice that P in this example is the set of vertices: 10, 13, 16, 2, 8, 20, 26, 6, and 18 in Figure 3.7b which assume that only nodes placed on adjacent vertices are connected. Now we solve the SDP in 3.18 for this specific example. As a result, the highest two values of λ_2

Table 3.3 Parameters of the Simulated WSN

PARAMETER	VALUE	PARAMETER	VALUE
RLT	20 (round)	N_{SPRN}	0–60 (relay node)
Total grid units	98	PNF	0%–60%
PDN	0%–60%	N_{CH}	20

(i.e., network connectivity with constraints on cost and lifetime) are associated with vertices 10 and 26.

By allocating the two SPRNs at these two vertices, we guarantee the network lifetime to be at least 20 rounds in addition to maximizing the backbone connectivity. For instance, we can see how removal of a single node, such as FPRNs at vertex 4 or 14 in Figure 3.7b, can cause a network partition. While using the SPRNs allocated at vertices 10 and 26, removal of at least two nodes is required to cause the network partition.   ∎

Finally, based on the output of Algorithms 3.1 and 3.5, locations of relay nodes (FPRNs and SPRNs) are determined optimally in terms of maximum connectivity and limited cost budget in addition to practical lifetime considerations. This can be easily proven given that the two solutions achieved by Algorithms 3.1 and 3.5 are optimal and independent. We assert that this two-phase solution can be easily extended to consider other constraints such as coverage, data fidelity, fault-tolerance, etc. It is also important to notice that Algorithms 3.1 and 3.5 are computationally efficient in practice with complexity of $O(n)$.

3.4 Performance Evaluation

In this section, we evaluate the performance of our proposed O3DwLC strategy under harsh environmental circumstances, where numerous probabilities of node failure and isolation are considered and 3-D setup is required. We compare our strategy to an efficient deployment strategy, called the shortest path 3-D grid deployment (SP3D). The SP3D strategy is usually used in environmental applications such as forest fire detection and soil experiments [29,31]. Moreover, SP3D strategy is used as a baseline in this research due to its efficiency in maintaining a predefined lifetime and choosing the minimum number of relay nodes required in constructing the network backbone. In SP3D, Algorithm 3.1 is used to construct the network backbone by allocating the minimum number of relay nodes on 3-D grid vertices. These relay nodes connect the pre-allocated cluster heads with the base station. Then, extra relay nodes are densely distributed near the network backbone devices in order to enhance their connectivity. Both O3DwLC and SP3D strategies are evaluated and compared using three different metrics:

1. *Backbone connectivity (λ_2)*: This criterion reflects deployed network reliability under harsh environmental characteristics and the ability to prolong lifetime. It indicates efficiency of the designed wireless sensor network.
2. *Number of CHs/RNs*: This indicates the system cost-effectiveness in harsh environments.
3. *Number of rounds*: This is a measurement of the total rounds for which the deployed network can stay operational. It reflects the efficiency of the estimated wireless sensor network lifetime.

Two main parameters are used in this comparison: probability of node failure (PNF) and Probability of disconnected (isolated) nodes (PDN). PNF is the probability of physical damage for the deployed node. PDN is the probability of a node to be disconnected while it still has enough energy to communicate with the base station. We chose these parameters as they are key factors in reflecting harshness of the monitored site in terms of weak signal reception and physical node damage.

3.4.1 Simulation Model

The O3DwLC and SP3D strategies are executed on 500 randomly generated WSN hierarchical graph topologies in order to get statistically stable results. For each topology, we apply a random node/link failure; performance metrics are computed accordingly. Dimensions of the 3-D deployment space are $700\times700\times200$ (m^3). Twenty irredundant cluster heads (i.e., responsible for different sensor nodes) in addition to one base station are randomly placed on 3-D cubic grid vertices using a Linear Congruential random number generator. We assume a predefined fixed time schedule for traffic generation at the cluster heads. Positions of relay nodes are found by applying the O3DwLC and the SP3D deployment strategies. We assume that each wireless sensor network is required to be operational for at least 20 rounds (lifetime constraint) using at most 60 relay nodes (cost constraint).

Based on experimental measurements taken in a site of dense trees [28], we set our system model variables to be as described in Tables 3.1 and 3.3. However, the n_c variable, which represent the search space size for the optimization problem in 3.18, will have different values varying from one deployment to another based on the locations of the backbone nodes reached by Algorithm 3.1 and their probabilistic communication ranges. Thus, the formulated sets of CGUs and MCGUs will vary from one deployment instance to another after applying Algorithms 3.2 through 3.4 on the resulting network backbone in the first phase of the O3DwLC strategy. Similarly, the values of N_{MST} and IRS are assigned based on the results of the first phase deployment; consequently, they vary from one instance to another. We assume fixed and equal transmission ranges to simplify the presentation of results in addition to applying identical grid edge lengths (=100 m). Nevertheless, the same simulation parameters are applicable for different transmission ranges varying from one device to another with unequal grid edge lengths. For comparison purposes, we simulate the lifetime and connectivity of different relay node counts and varying PNF and PDN (0%–60%) for both strategies, O3DwLC and SP3D. As we mentioned previously, each simulation experiment is repeated 500 times and the average results hold a confidence interval of no more than 5% of the average (over 500 runs) at a 95% confidence level.

3.4.2 Simulation Results

While the O3DwLC strategy optimizes locations of second phase relay nodes in order to achieve the highest connectivity (λ_2), SP3D attempts to find locations of extra relay nodes (deployed after constructing the optimized backbone) by distributing them on 3-D grid vertices that are more likely to increase the network connectivity as we described above. It is expected from O3DwLC to outperform the SP3D in terms of connectivity and total number of nodes as shown in Figure 3.10. Figure 3.10 presents the average λ_2 for both strategies using different totals of RNs, where the number of cluster heads is fixed to 20 to see the effect of relay node placement and PDN=0.2. It is clear how an increment in the deployed RN leads to an increment in connectivity even in the presence of 20% disconnected nodes using O3DwLC strategy. Moreover, Figure 3.10 shows how efficient the networks generated by O3DwLC are in terms of the utilized RN count (and thus, the cost) to achieve a specific connectivity

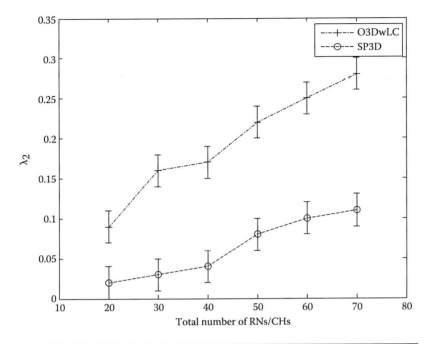

Figure 3.10 Connectivity vs. the deployed nodes' count.

requirement. For instance, using 30 nodes only, O3DwLC strategy achieves a connectivity value higher than the connectivity value achieved by SP3D using 70 nodes. This indicates higher savings in terms of the network cost.

We also investigate the effect of PDN, which varies from 0% to 60%, on lifetime using both strategies as shown in Figure 3.11. We select the total number of nodes in this comparison to be 40 (20 CHs and 20 RNs) to avoid extreme cases where total number of RNs is too high or too low. From Figure 3.11, we can see how the WSNs generated by O3DwLC can stay operational for longer time than the SP3D which indicates high reliability. This happens where there exists at least one operational path from each cluster head to the base station even in the presence of 60% PDN [1]. From Figures 3.10 and 3.12 we also observe that increasing the connectivity value of the WSN will increase its lifetime since both figures are simulating the same WSNs with the same total number of nodes.

In Figure 3.12, we examine the effect of the lifetime constraint on the O3DwLC and SP3D strategies when PDN=20%. Under that PDN, O3DwLC is still much better in terms of the total rounds for

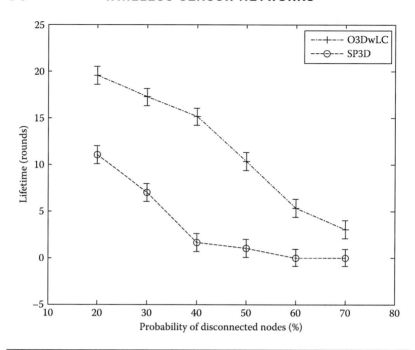

Figure 3.11   Lifetime vs. PDN.

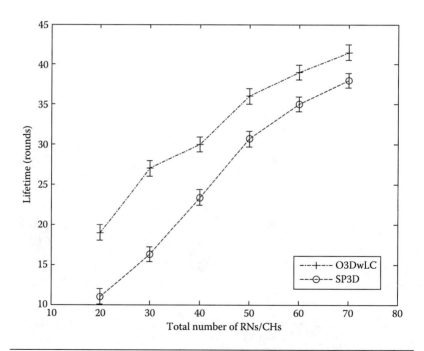

Figure 3.12   Lifetime vs. number of nodes under PDN=0.2 and RLT=20.

which a network can stay operational. This is a very attractive feature in outdoor monitoring. Not surprisingly, the difference in lifetime of the WSNs generated using both strategies decrease as the total number of relay nodes increases due to node density increment. This makes the deployed networks tightly connected and harder to partition.

Figures 3.13 and 3.14 show how O3DwLC strategy outperforms SP3D strategy under different PDN and PNF values, respectively. This supports our O3DwLC deployment strategy efficiency in terms of connectivity. Wireless sensor networks generated by O3DwLC strategy stay connected even when PDN = PNF = 50%. This is another desired and required performance issue in harsh outdoor environmental applications. However, as the PNF/PDN values achieve a specific level where the available number of functional nodes cannot tolerate the failure, the WSN connectivity decreases dramatically and network partition occurs. This explains the prominent degradation in WSN connectivity even when O3DwLC strategy is utilized.

We note that choosing an appropriate value of N_{SPRN} is highly dependent on the probability of node failure. Figure 3.15 shows the effect of PNF on the choice of N_{SPRN}. For low values of PNF, only few

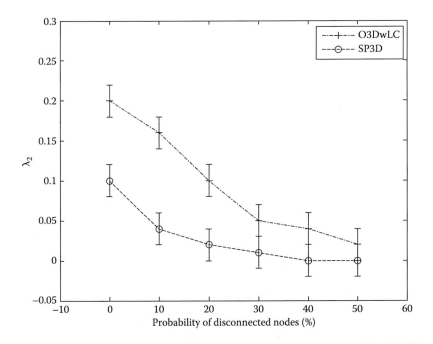

Figure 3.13   Connectivity vs. probability of disconnected CHs/RNs.

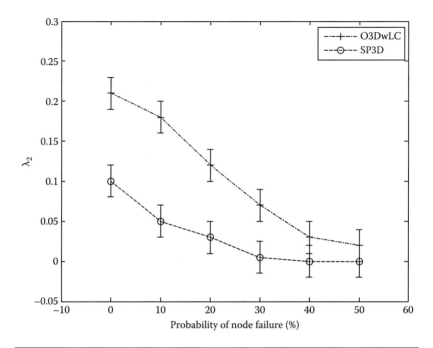

Figure 3.14   WSN connectivity vs. probability of CHs/RNs failure.

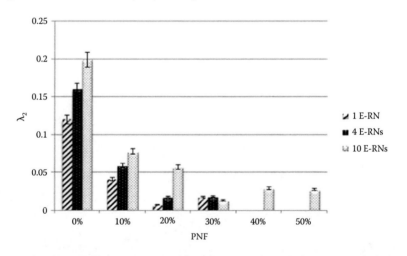

Figure 3.15   Indicating SPRNs count required to overcome specific PNF value.

extra relay nodes (E-RNs) are needed. On the other hand, at least 10 E-RNs are needed to guarantee connectivity in environments with a 50% PNF. Considering such percentage of failure in the monitored site during the early stages of the deployment plan would have a great effect on the network performance in practice.

3.5 Conclusion

In this chapter, we explored the problem of relay deployment in WSNs applied in 3-D outdoor environmental applications; we aimed at maximizing network connectivity with constraints on lifetime and cost. Such deployment problems have been shown to be NP-hard. Finding near optimal solutions is also an NP-hard. To address this complexity, we propose an efficient two-phase relay node deployment in 3-D space using minimum spanning tree and semi-definite programming. The first phase is using the minimum spanning tree to set up a connected network backbone with the minimum number of relay nodes for cost efficiency. In the second phase, we found a set of a relatively small count of candidate positions. And we used the semi-definite programming to optimize the relay node placement on these positions to achieve the maximum backbone connectivity for guaranteed lifetime period within a limited cost budget. Toward a more practical solution, application-specific signal propagation and lifetime models were considered in addition to limiting the huge search space of the targeted deployment problem.

The signal propagation model provided more realistic communication properties between the deployed nodes in order to precisely describe their ability to communicate with each other. As for the lifetime model, several lifetime definitions in the literature were discussed and compared based on practical metrics and parameters to reflect the appropriateness of these definitions in environmental applications. The extensive simulation results, obtained under harsh operational conditions, indicated that the proposed two-phase strategy can provide tightly-connected networks and practically-guaranteed lifetime for environmental applications. Moreover, deployment strategy and results presented in this chapter can provide a tangible guide for network provisioning in large-scale environmental applications which require 3-D setups. In addition, they are applicable for different grid shapes and environment characteristics (e.g., various signal attenuation and path loss levels).

Future work would investigate optimal deployment problem-solving in further environment monitoring scenarios, where a subset of the relay nodes may have the mobility feature to repair connectivity and prolong network lifetime. Also, of practical interest is the node placement under varying transmission range and/or different power supply from one node to another for more energy-efficient solutions.

References

1. A. Abbasi, M. Younis, and K. Akkaya, Movement-assisted connectivity restoration in wireless sensor and actor networks, *IEEE Transactions on Parallel Distributed Systems*, vol. 20, no. 9, pp. 1366–1379, 2009.
2. H. M. F. AboElFotoh, S. S. Iyengar, and K. Chakrabarty, Computing reliability and message delay for cooperative wireless distributed sensor networks subject to random failures, *IEEE Transactions on Reliability*, vol. 54, no. 1, pp. 145–155, 2005.
3. I. Akyildiz, W. Su, Y. Sankarasubramaniam, and E. Cayirci, A survey on sensor networks, *IEEE Communications Magazine*, vol. 40, no. 8, pp. 102–114, 2002.
4. S. N. Alam and Z. Haas, Coverage and connectivity in three-dimensional networks, In *Proceedings of the ACM International Conference on Mobile Computing and Networking (MobiCom)*, Los Angeles, CA, 2006, pp. 346–357.
5. A. Bari, A. Jaekel, and S. Bandyopadhyay, Optimal placement of relay nodes in two-tiered, fault tolerant sensor networks, In *Proceedings of the IEEE International Conference on Computers and Communications (ICCC)*, Aveiro, Portugal, 2007, pp. 159–164.
6. M. Bhardwaj, A. Chandrakasan, and T. Garnett, Upper bounds on the lifetime of sensor networks, In *Proceedings of the IEEE International Conference on Communications (ICC)*, St. Petersburg, 2001, pp. 785–790.
7. S. Boyd, Convex optimization of graph Laplacian eigenvalues, *Proceedings of the International Congress of Mathematicians*, vol. 3, no. 63, pp. 1311–1319, 2006.
8. A. Cerpa and D. Estrin, Ascent: Adaptive self-configuring sensor networks topologies, *IEEE Transactions on Mobile Computing*, vol. 3, no. 3, pp. 272–285, 2004.
9. Y. Chen and Q. Zhao, On the lifetime of wireless sensor networks, *IEEE Communications Letters*, vol. 9, no. 11, pp. 976–978, 2005.
10. P. Cheng, C. Chuah, and X. Liu, Energy-aware node placement in wireless sensor networks, In *Proceedings of the IEEE Global Telecommunications Conference (GLOBECOM)*, Dallas, TX, 2004, pp. 3210–3214.
11. X. Cheng, D. Du, L. Wang, and B. Xu, Relay sensor placement in wireless sensor networks, *Journal of Wireless Networks*, vol. 14, no. 3, pp. 347–355, 2008.
12. C. Decayeux and D. Seme, A new model for 3-D cellular mobile networks, In *Proceedings of the International Symposium Parallel and Distributed Computing (ISPDC)*, Cork, Ireland, 2004.
13. A. Efrat, S. Fekete, P. Gaddehosur, J. Mitchell, V. Polishchuk, and J. Suomela, Improved approximation algorithms for relay placement, In *Proceedings of the 16th Annual European Symposium on Algorithms*, Karlsruhe, Germany, 2008, pp. 356–367.
14. A. Ghosh and S. Boyd, Growing well-connected graphs, In *Proceedings of the IEEE Conference on Decision and Control*, San Diego, CA, 2006, pp. 6605–6611.

15. L. Hoesel, T. Nieberg, J. Wu, and P. Havinga, Prolonging the lifetime of wireless sensor networks by cross layer interaction, *IEEE Transactions on Wireless Communications*, vol. 11, no. 6, pp. 78–86, 2004.

16. Y. Hou, Y. Shi, H. Sherali, and S. F. Midkiff, On energy provisioning and relay node placement for wireless sensor network, *IEEE Transactions on Wireless Communications*, vol. 4, no. 5, pp. 2579–2590, 2005.

17. M. Ibnkahla, ed., *Adaptation and Cross Layer Design in Wireless Networks*, CRC Press, Boca Raton, FL, 2008.

18. M. Ishizuka and M. Aida, Performance study of node placement in sensor networks, In *Proceedings of the International Conference on Distributed Computing Systems Workshops (ICDCSW)*, Tokyo, Japan, 2004, pp. 598–603.

19. A. Kashyap, S. Khuller, and M. Shayman, Relay placement for higher order connectivity in wireless sensor networks, In *Proceedings of the IEEE Conference on Computer Communications (INFOCOM)*, Barcelona, Spain, 2006, pp. 1–12.

20. S. Lee and M. Younis, Optimized relay placement to federate segments in wireless sensor networks, *IEEE Transactions on Selected Areas in Communications*, vol. 28, no. 5, pp. 742–752, 2010.

21. J. Lee, T. Kwon, and J. Song, Group connectivity model for industrial wireless sensor networks, *IEEE Transactions on Industrial Electronics*, vol. 57, no. 5, pp. 1835–1844, 2010.

22. N. Li and J. C. Hou, Improving connectivity of wireless ad hoc networks, In *Proceedings of the IEEE International Conference on Mobile and Ubiquitous Systems: Networking and Services (MobiQuitous)*, San Diego, CA, 2005, pp. 314–324.

23. H. Liu, P. Wan, and X. Jia, Fault-tolerant relay node placement in wireless sensor networks, *Lecture Notes in Computer Science (LNCS)*, vol. 3595, pp. 230–239, 2005.

24. E. Lloyd and G. Xue, Relay node placement in wireless sensor networks, *IEEE Transactions on Computers*, vol. 56, no. 1, pp. 134–138, 2007.

25. S. Misra, S. Hong, G. Xue, and J. Tang, Constrained relay node placement in wireless sensor networks to meet connectivity and survivability requirements, In *Proceedings of the IEEE Conference on Computer Communications (INFOCOM)*, Phoenix, AZ, 2008, pp. 281–285.

26. C. Ortiz, J. Puig, C. Palau, and M. Esteve, 3D wireless sensor network modeling and simulation, In *Proceedings of the IEEE Conference on Sensor Technologies and Applications (SensorComm)*, Valencia, Spain, 2007, pp. 307–312.

27. V. Ravelomanana, Extremal properties of three-dimensional sensor networks with applications, *IEEE Transactions on Mobile Computing*, vol. 3, no. 3, pp. 246–257, 2004.

28. J. Rodrigues, S. Fraiha, H. Gomes, G. Cavalcante, A. de Freitas, and G. de Carvalho, Channel propagation model for mobile network project in densely arboreous environments, *Journal of Microwaves and Optoelectronics*, vol. 6, no. 1, pp. 189–206, 2007.

29. A. Singh, M. A. Batalin, V. Chen, M. Stealey, B. Jordan, J. C. Fisher, T. C. Harmon, M. H. Hansen, and W. J. Kaiser, Autonomous robotic sensing experiments at San Joaquin River, In *Proceedings of the IEEE International Conference on Robotics and Automation (ICRA)*, Rome, Italy, 2007, pp. 4987–4993.
30. W. Song, R. Huang, M. Xu, B. Shirazi, and R. LaHusen, Design and deployment of sensor network for real-time high-fidelity volcano monitoring, *IEEE Transactions on Parallel and Distributed Systems*, vol. 21, no. 11, pp. 1658–1674, 2010.
31. B. Son, Y. Her, and J. Kim, A design and implementation of forest-fires surveillance system based on wireless sensor networks for South Korea mountains, *International Journal of Computer Science and Network Security*, vol. 6, no. 9, pp. 124–130, 2006.
32. H. Tan, Maximizing network lifetime in energy-constrained wireless sensor network, In *Proceedings of the ACM International Wireless Communications and Mobile Computing Conference (IWCMC)*, Vancouver, BC, 2006, pp. 1091–1096.
33. J. Tang, B. Hao, and A. Sen, Relay node placement in large scale wireless sensor networks, *Computer Communication*, vol. 29, no. 4, pp. 490–501, 2006.
34. G. Tolle, J. Polastre, R. Szewczyk, and D. Culler, A macroscope in the redwoods, *In Proceedings of the ACM Conference on Embedded Networked Sensor Systems (SenSys)*, San Diego, CA, 2005, pp. 51–63.
35. K. Xu, H. Hassanein, G. Takahara, and Q. Wang, Relay node deployment strategies in heterogeneous wireless sensor networks, *IEEE Transactions on Mobile Computing*, vol. 9, no. 2, pp. 145–159, 2010.
36. W. Ye, J. Heidemann, and D. Estrin, Medium access control with coordinated adaptive sleeping for wireless sensor networks, *IEEE/ACM Transactions on Networking (ToN)*, vol. 12, no. 3, pp. 493–506, 2004.
37. M. Younis and K. Akkaya, Strategies and techniques for node placement in wireless sensor networks: A survey, *Elsevier Ad Hoc Network Journal*, vol. 6, no. 4, pp. 621–655, 2008.

4

OPTIMIZED RELAY PLACEMENT FOR WIRELESS SENSOR NETWORKS FEDERATION IN ENVIRONMENTAL APPLICATIONS*

Advanced sensing technologies have enabled the wide use of Wireless Sensor Networks (WSNs) in large-scale Outdoor Environment Monitoring (OEM) [1,2]. The most notable among these applications are those in harsh environments, such as forest fire and flood detection applications [1,3]. WSNs in such applications are not only subject to severe damages that might partition the network into disjointed sectors as shown in Figure 4.1, but also can work together in detecting and preventing significant disasters that threaten the environment we are living in (see Figure 4.2) [4]. To enable their connection and interaction, disjointed WSN sectors need to be (and to stay) within reach of each other in the presence of high Probabilities of Node Failure (PNF) and Probabilities of Link Failure (PLF); thus, the connectivity has a significant impact on the effectiveness of federated WSNs in OEM.

In general, connectivity problems can be dealt with either by populating relay nodes or by utilizing mobile nodes [5,6]. For example, in Reference [5], the lowest number of relays is added to a disconnected static WSN so the network remains connected. In Reference [6], mobile nodes are used to address k-connectivity requirements, where k is equal to 1 and 2. The idea is to identify the least relay node count

---

* This chapter has been coauthored with Hossam S. Hassanein, Waleed M. Alsalih, and Mohamad Ibnkahla.

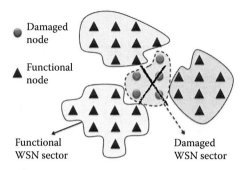

Figure 4.1   Vast damage partitioning a WSN into disjointed sectors.

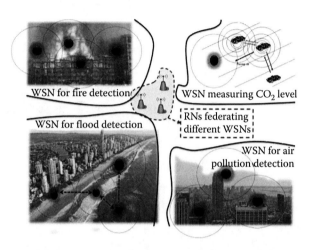

Figure 4.2   WSNs collaborating in protecting our environment.

that should be repositioned in order to re-establish a particular level of connectivity. However, connecting WSN sectors in OEM is more challenging due to expensive relays and the huge distances separating different sectors which might exceed twice the communication range of a relay node. In this chapter, we investigate an efficient way for relay node placement which addresses the preceding challenges in OEM applications.

This node placement problem has been shown in Reference [7] to be NP-hard. Finding near optimal approximate solutions is also NP-hard in some cases. To address this complexity, we propose an Optimized two-phase Relay Placement (ORP) approach. The first phase sets up a connected network backbone using a reasonably small number of relays, which we call First Phase Relay Nodes (FPRNs). The first phase also finds a set of candidate locations for relays that are deployed in the second phase, which we call Second Phase Relay Nodes (SPRNs). The second phase aims at deploying the available number of SPRNs in the candidate positions obtained from the first phase in such a way that maximizes the WSN connectivity. The two schemes we present in this chapter differ in the first phase. The first scheme is the Grid-based Optimized Relay Placement (GORP) in which all relays are assumed to be deployed on grid vertices as shown in Figure 4.3. The second scheme is a general Optimized Relay Placement (ORP) scheme where relays may be placed at any point in the field. The general ORP scheme utilizes some interesting geometrical structures to connect the disjointed sectors and to find candidate positions for the SPRNs. Once the candidate positions for SPRNs are found, selecting the locations of the SPRNs is done by formulating the problem as a relaxed Semi-Definite Program (SDP) and solving it using a standard SDP solver.

Major contributions of this chapter can be described as follows. We formulate a generic relay node placement problem for maximizing connectivity with constraints on the relay count. We propose two optimized schemes for the deployment problem. Performance of the proposed schemes is evaluated and compared to other existing approaches in the literature.

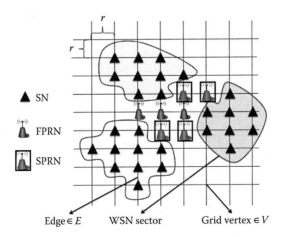

Figure 4.3 Grid-based network architecture.

The remainder of this chapter is organized as follows. In Section 4.1, related work is surveyed. In Section 4.2, our two-phase deployment strategy is described. The performance of the proposed strategy is evaluated and compared to other deployment strategies in Section 4.3. Section 4.4 concludes the chapter.

4.1 Related Work

In Reference [8], Lloyd and Xue opt to deploy the fewest RNs such that each sensor is connected to at least one RN, and the inter-RN network is strongly linked by forming a Minimum Spanning Tree (MST) and employing a Geometric Disk Cover algorithm. While in Reference [9], the authors solve a Steiner tree problem to deploy the fewest RNs. Although the MST and the Steiner tree may guarantee the lowest cost by occupying the minimum number of relays, they tend to establish an inefficient WSN topology in terms of connectivity, as discussed in Section 4.4.

Unlike [8,9], Xu et al. [10] study a random RN deployment that considers the network connectivity for the longest WSN operational time. The authors proposed an efficient WSN deployment that maximizes the network lifetime when RNs communicate directly with the Base Station (BS). In this study, it was established that different energy consumption rates at different distances from the BS render uniform RN deployment a poor candidate for network lifetime extension. Alternatively, a weighted random deployment is proposed. In this random deployment, the density of RNs' deployment is increased as the distance to the BS increases; thus, distant RNs can split their traffic among themselves. This in turn extends the average RN's lifetime while maintaining a connected WSN.

Furthermore, the approach presented in Reference [11] aims at considering WSN connectivity in harsh environments. It counters faulty nodes causing connectivity problems by repositioning pre-identified spare relays from different parts of a 2-D grid model. The grid is divided into cells. Each cell has a head that advertises the available spare nodes in its cell or requests the spares for its cell. A quorum-based solution is proposed to detect the intersection of the requests within the grid. Once the spares are located, they are moved to a cell with failed nodes.

In Reference [12], a distributed recovery algorithm is developed to address specific connectivity degree requirements. The idea is to identify the least set of nodes that should be repositioned in order to reestablish a particular level of connectivity. Nevertheless, these references (i.e., [10–12]) do not minimize the relay count, which may not be cost-effective in environmental monitoring applications. Consequently, considering both connectivity and relay count is the goal of References [13,14]. In Reference [13], Lee and Younis focus on designing an optimized approach for federating disjointed WSN segments (sectors) by populating the least number of relays. The deployment area is modeled as a grid with equal-sized cells. The optimization problem is then mapped to selecting the fewest count of cells to populate relay nodes such that all sectors are connected. In an earlier work [14], we proposed an Integer Linear Program (ILP) optimization problem to determine sensors and relay positions on grid vertices that maximize the network lifetime while maintaining the k-connectivity level.

Unlike [13,14], in this chapter, ORP considers the network connectivity and the relay count in a different way. Bearing in mind that the disjointed sectors and the minimum number of RNs required to join them represent the WSN backbone, ORP aims to maximize the backbone connectivity by placing a limited number of extra relays. This in turn renders more sustainable WSN topologies in harsh environments than those generated by References [8,9], and unlike [10–12], ORP addresses the network connectivity problems without violating its cost-effectiveness.

4.2 Optimized WSN Federation

In this section, we present our scheme for federating disconnected WSN sectors with the purpose of maximizing algebraic connectivity.

4.2.1 Definitions and Assumptions

A WSN sector is a set of connected relaying nodes which we call sector nodes (SNs). The exact location of each SN is assumed to be known in advance. Each WSN sector is represented using a virtual super single node (SSN). The x-coordinate (y-coordinate) of an SSN of a particular sector is the average of the maximum and the minimum x-coordinates (y-coordinates) of SNs in that sector.

An edge connecting two SNs from two different sectors is said to *connect* the two sectors. The distance between two sectors is the length of the shortest edge connecting them.

The transmission range of all relaying nodes is modeled as a circle with a radius of r m (i.e., identical transmission ranges).

Now, the problem can be defined as follows:

Given a set of WSN sectors along with the locations of their SNs, determine the locations of Q Relaying Nodes (RNs) so that connectivity among WSN sectors is established and maximized.

The network is modeled as a graph $G = (V, E)$, where V is the set of all RNs and SSNs. E is the set of edges connecting SSNs and RNs. An SSN shares an edge with an RN if the RN is within the transmission range of at least one SN belonging to the SSN's sector.

4.2.2 Deployment Strategy

The node placement problem addressed in this chapter has an infinite search space; this is because each RN may be placed at any point in a 2-D plane. We propose two schemes to restrict the search space to a finite number of locations and to make the optimization problem discrete. The first scheme is the grid deployment in which locations of RNs are limited to the vertices of a grid as shown in Figure 4.3. The second scheme constructs a set of edges connecting WSN sectors, and locations of new RNs are limited to a set of points along those edges. Those edges are derived from the Delaunay Triangulation [15] and the Steiner tree [16] of the virtual SSNs.

4.2.2.1 Grid-Based ORP (GORP)   This scheme assumes that SNs and RNs are placed on the vertices of a grid whose edges have a length equal to the transmission range of SNs and RNs [i.e., r (m)]. This grid architecture is shown in Figure 4.3. In this scheme, RNs are deployed in two phases.

4.2.2.1.1 First phase deployment   RNs deployed in the first phase are called first phase RNs (FPRNs). The purposes of the first phase are to deploy the minimum number of FPRNs to federate WSN sectors (i.e., SNs of all WSN sectors and FPRNs form a connected graph), and to construct a finite set of potential locations for the second phase RNs (SPRNs). FPRNs' positions are determined using the Minimum

Spanning Tree (MST) of the SSNs. We find the MST of a complete graph whose vertices are the SSNs and the weight of each edge is the distance between the two sectors it connects. If two SSNs share an edge, we deploy the minimum number of FPRNs to connect them on the grid. If S_i and S_j are two WSN sectors, let $Connect_G(S_i, S_j)$ denote the smallest cardinality set of grid vertices that form a path from an SN in S_i to an SN in S_j. If S_i and S_j share an edge in the MST, $|Connect_G(S_i, S_j)|$ FPRNs are deployed to connect them. In other words, we deploy the minimum number of RNs on grid vertices to facilitate the communication between the two WSN sectors. Algorithm 4.1 presents a high level description of the first phase.

Algorithm 4.1: Grid-based deployment of FPRNs

Function Phase1 (S, N)

Input:
    S: The set of all SSNs.
    N: The set of all SNs (their locations and the sectors they belong to).

Output:
    F: A set of grid vertices where FPRNs are located so that all WSN sectors are connected.
    Gr: The set of all candidate positions for SPRNs.

begin
    $F = \phi$;
    Gr = the set of all grid vertices;
    Find MST(S) which is the MST of a complete graph whose vertices are S and the weight of each edge is the distance between the two sectors it connects;

foreach edge e in MST(S) **do**
    $F = F \cup Connect_G(S_i, S_j)$, where e is connecting the SSN of S_i and that of S_j;
    $Gr = Gr - Connect(S_i, S_j)$;

    end
    Return (F, Gr);

End

4.2.2.1.2 Second Phase Deployment   In the second phase, we formulate and solve an optimized SDP with the objective function of maximizing the algebraic connectivity without exceeding a specific budget for the total number of SPRNs. The outcome of the first phase is a connected graph, which we denote by B, whose vertices are the SSNs and the FPRNs. It is a backbone that makes the whole network connected. Connectivity of B is measured by the second smallest eigenvalue λ_2 of the Laplacian matrix $L(B)$ [17]. The Laplacian matrix is a 2-D matrix that has -1 in the element (i,j), if there is a connection between nodes i and j and 0 otherwise. It has the degree of node i in the element (i,i) (see Figure 4.4). Given $L(B)$, the algebraic connectivity of B is the second smallest eigenvalue λ_2. By increasing the value of λ_2 in $L(B)$, we tend to increase the required number of nodes and communication links to disjoint (disconnect) B. This is because of the proportional relationship between the value of λ_2 and the number of nodes/links which can cause network partitions as shown in Figure 4.4. A better connectivity improves the ability of the network to overcome significant topology changes caused by communication quality changes and node failures. This is achieved through deploying extra RNs, which are called the second phase relay nodes (SPRNs), in the second phase of our deployment strategy. Deploying extra RNs means adding more nodes and edges to the graph B.

The objective in the second phase is to find the best positions for SPRNs, out of all grid vertices, such that λ_2 of the resulting backbone

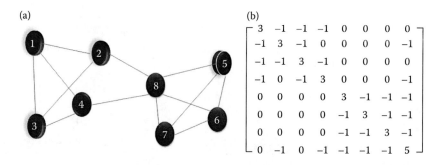

Figure 4.4   (a) A graph with 8 nodes and 13 links. The graph's connectivity characteristics are: one node to disconnect (removal of node 8), two links to disconnect (removal of links connecting node 8 to nodes 2 and 4). A Laplacian matrix of this graph is shown in (b) and λ_2 of the matrix in (b) is equal to 0.6277. As λ_2 increases the node/link count required to partition the network increases.

graph is maximized with constraints on the total number of SPRNs to be deployed. Let N_{SPRN} denote the total number of SPRNs to be deployed. Assume we have n_c grid vertices as candidate positions for SPRNs. We want to choose the optimal N_{SPRN} grid vertices amongst these n_c vertices. We can then formulate this optimization problem, with reference to Table 4.1, as

$$\max \lambda_2 \left(L(\alpha) \right)$$

$$s.t. \sum_{i=1}^{n_c} \alpha_i = N_{SPRN}, \alpha_i \in \{0, 1\}, \tag{4.1}$$

where,

$$L(\alpha) = L_i + \sum_{i=1}^{n_c} \alpha_i A_i A_i^T \tag{4.2}$$

However, an exhaustive search scheme is required to find the optimal solution, which is computationally expensive as it takes exponential time to find the best solution out of $\begin{pmatrix} nc \\ N_{SPRN} \end{pmatrix}$ possible solutions; this is besides the cost of finding λ_2 for each solution. Therefore, we need a computationally efficient means to find near optimal solutions to this optimization problem. For that purpose, we reformulate the problem as a standard Semi-definite Program (SDP) optimization problem [17,18], which can be solved using any standard SDP solver.

Table 4.1   Notations Used in the Placement Problem

NOTATION	DESCRIPTION
α_i	A binary variable equals 1 when RN at vertex i in the 3-D grid is allocated and 0 otherwise.
A_i	Incidence matrix that results by adding RN_i in the 3-D grid; $A_i = [a_1, a_2, ..., a_m]$, where a_i is the vector that consists of n elements that can take a value of either 0, 1 or −1 and m is the total number of edges that is produced by adding RN_i. For example, if adding RN_i will establish a connection between node 1 and 3, then 1st element is set to 1 and 3rd element is set to −1 and all of remaining elements are set to zeros.
N	Which is the total of FPRNs, and BS.
L_i	Initial Laplacian matrix produced by the allocated FPRNs.
I_{nxn}	Identity matrix of size n by n.

However, we need to relax the Boolean constraint $\alpha \in \{0, 1\}$ to be a linear constraint $\alpha \in [0, 1]$ as follows:

$$\max \lambda_2 \left(L(\alpha) \right)$$

$$s.t. \ \sum_{i=1}^{n_c} \alpha_i = N_{SPRN}, \ 0 \leq \alpha_i \leq 1, \tag{4.3}$$

We observe that the optimal value of the relaxed problem in 4.3 gives an upper bound for the optimal value of the original optimization problem in 4.1. The optimal solution for 4.3 is obtained numerically using one of the standard SDP solvers (e.g., the SDPA-M software package). Finally, we use a heuristic to obtain a Boolean vector from the SDP optimal solution as a solution for the original problem in 4.1. In this article, we consider a simple heuristic, which is to set the largest $N_{SPRN} \alpha_i$'s to 1 and the rest to 0.

The optimization problem in 4.3 is convex with linear constraints [18]. Thereby we introduce the following Theorem.

Theorem 4.1: The optimization problem in 4.3 is mathematically equivalent to the following SDP optimization problem

$$\max S$$

$$s.t. \ S\left(I_{nxn} - \frac{1}{n} 11^T \right) \preccurlyeq L(\alpha), \sum_{i=1}^{n_c} \alpha_i = N_{SPRN}, \ 0 \leq \alpha_i \leq 1, \tag{4.4}$$

Where S is a scalar variable and $\preccurlyeq$ denotes the positive semi-definiteness (i.e., all eigenvalues of the matrix are greater than or equal to zero).

Proof: Let $V \in \mathbf{R}^n$ be the corresponding eigenvector of $\lambda_2(L(\alpha))$. Thus, $1^T V = 0$, and $V = 1$.

Since,

$$L(\alpha)V = \lambda_2 V \tag{4.5}$$

Hence,

$$V^T L(\alpha)V = \lambda_2 V^T V = \lambda_2 \tag{4.6}$$

$$\Rightarrow \lambda_2 \left(L(\alpha) \right) = \inf_V \{ V^T L(\alpha)V \, | \, 1^T V = 0, \ \text{and} \ V = 1 \} \tag{4.7}$$

Let,

$$L'(\alpha) = L(\alpha) - S\left(I_{nxn} - \frac{1}{n}11^T\right) \tag{4.8}$$

Thus for any $V_{n\times1}$ where $1^T V = 0$, and $V = 1$, we get

$$V^T L'(\alpha)V = V^T L(\alpha)V - S\left(V^T I_{nxn}V - \frac{1}{n}\left(V^T 1\right)\left(1^T V\right)\right)$$

$$= V^T L(\alpha)V - S \tag{4.9}$$

Hence, for $L'(\alpha)$ to be positive semi-definite, the maximum value of S should be

$$S = inf_V \{V^T L(\alpha)V \,|\, 1^T V = 0, \text{ and } V = 1\}, \tag{4.10}$$

From Equations 4.12 and 4.15,

$$S = \lambda_2\left(L(\alpha)\right) \tag{4.11}$$

Therefore, maximizing S in Equation 4.4 is equivalent to maximizing $\lambda_2\left(L(\alpha)\right)$ in Equation 4.3 if the constraints are satisfied.

∎

Example 4.2: Assume we have up to two extra relay nodes (SPRNs) to maximize connectivity of the backbone generated in first phase. In this case, $N_{SPRN} = 2$. We start by calculating the initial Laplacian matrix L_i associated with the backbone to be used in Equation 4.2. With reference to Table 4.1, we set α_i to 1 and calculate A_i for each vertex i on the grid. Now we solve the SDP in Equation 4.4 for this specific example. As a result, candidate positions with the highest two values of λ_2 will be chosen to allocate the available two SPRNs. ∎

Algorithm 4.2 presents a high level description of the second phase.

Algorithm 4.2: SPRNs deployment

Function SPRNs (B, Z)

Input:
    B: A connected graph of the SSNs and FPRNs.
    $Z = \{z_1, z_2, z_3, \ldots, z_{|Z|}\}$: The set of all candidate positions for SPRNs
    N_{SPRN}: The total number of SPRNs.

Output:
F: A set of points where SPRNs are located.

begin
   $\mathbf{L}_i$ = Laplacian matrix of B;
   for $(i = 1; i < |Z|; i{+}{+})$
   $\mathbf{A}_i$ = coefficient matrix corresponding to the location z_i;
   end
   $\mathbf{SP}$ = Solution of SDP in Equation 4.2;
   Return SP;

End

4.2.2.2 The General Non-Grid (ORP)   While using grid vertices as candidate positions for SPRNs helps in discretizing the search space, it gives a relatively large number of candidate positions. This may affect both the time needed to solve an SDP and the quality of the obtained solution, especially in sparse networks.

In non-grid-based deployment, we try to construct a smaller set of candidate positions by selecting a set of points that have a geometrical property, which makes them more likely to be used as a location for SPRNs. This scheme constructs a set of edges connecting WSN sectors, and the search space of SPRN locations is limited to a set of points along those edges. We use the Delaunay Triangulation (DT) and the Steiner tree of the virtual SSNs to construct these edges. These two geometrical structures possess several nice properties that make them good sources of potential SPRN locations. The DT, for example, is a super-graph of both the Nearest Neighbor Graph (NNG) and the Euclidean MST. The Steiner tree also has a nice property of connecting a set of points (i.e., the SSNs here) with a network of edges with a minimum length. These properties seem to be useful in federating WSN sectors; it is intuitive that SPRNs will be used to connect sectors that are close to each other, hence the use of NNG edges. Meeting the limited budget of SPRNs requires using minimum length spanning edges, hence the use of Steiner tree and MST.

Grid-based and non-grid-based deployments differ in the first phase; i.e., they differ in the way they build an initial connected graph and in the set of candidate positions for SPRNs. However, the second phase is the same for both deployment strategies; they both solve an SDP to deploy SPRNs. The non-grid-based deployment uses the Steiner tree to deploy FPRNs. A Steiner tree of all SSNs is constructed, and FPRNs are deployed along Steiner tree points and edges. Before describing our non-grid-based deployment, we make some definitions. If e is an edge, $P(e, r)$ is the minimum cardinality set of points that partition e into smaller sectors of length at most r. For example, if the Euclidean length of e is 8 and $r = 2$, then $|P(e, r)| = 3$. If S_i and S_j are two WSN sectors, let $Connect(S_i, S_j) = P(e_{ij}, r)$ where e_{ij} is the shortest edge connecting an SN in S_i to an SN in S_j.

In our non-grid-based deployment, we build the Steiner tree of all SSNs, and we consider Steiner tree points to be a WSN sector with a single SN. Then, if two sectors S_i and S_j share an edge in the Steiner tree, $|Connect(S_i, S_j)|$ FPRNs are deployed to connect them. This makes a connected graph connecting all WSN sectors. We also use the DT to construct a set of candidate positions for SPRNs. Locations of SPRNs are limited to points along edges connecting SNs of WSN sectors that share a Delaunay edge. Algorithm 4.3 gives a high level description of the first phase of the non-grid-based deployment.

4.3 Performance Evaluation

4.3.1 Simulation Environment

Using MATLAB®, we simulate randomly generated WSNs which have the graph topology proposed in the previous section and consist of a varying number of partitioned sectors[*]. To solve the previously modeled SDP optimization problem, we used the SDPA-M MATLAB Package [19].

4.3.2 Performance Metrics and Parameters

To evaluate our ORP approach, we tracked the following performance metrics:

- *Connectivity* (λ_2): This criterion reflects the federated network reliability under harsh environmental characteristics. It gives an indication for the designed WSN efficiency.

---

[*] Random in size (of SNs) and positions.

- *Number of RNs (Q_{RN})*: This represents the cost-effectiveness of the deployment approach.

Four main parameters are used in the performance evaluation: (1) Probability of Node Failure (PNF), (2) Probability of Link Failure (PLF), (3) Number of SSNs (Q_{SSN}), and (4) Deployment Area (DA). PNF is the probability of physical damage for the deployed node. PLF is the probability of communication link failure due to bad channel conditions, which uniformly affects any of the network links. We chose these two parameters as they are key factors in reflecting harshness of the monitored site in terms of weak signal reception and physical node damage. As for the Q_{SSN}, it represents the degree of the network damage in case of partitioned WSNs and represents the problem complexity in case of federating multiple WSNs. And the DA reflects the scalability and applicability of the proposed deployment strategies in large-scale applications.

Algorithm 4.3: Non-grid-based deployment of FPRNs

Function Phase1 *(S, N)*

Input:
   S: The set of all SSNs.
   N: The set of all SNs (their locations and the sectors they belong to).

Output:
   F: A set of points where FPRNs are located so that all WSN sectors are connected.
   Gr: The set of all candidate positions for SPRNs.

begin
   $F = \phi$;
   $Gr = \phi$;
   Find ST(*S*) which is the Steiner tree for the set of points in *S*;

foreach edge *e* in ST(*S*) **do**
   $F = F \cup$ *Connect* (S_i, S_j), where *e* is connecting the SSN of S_i and that of S_j;

   end

Find DT(S) which is the DT of the points in S;

foreach edge e in DT(S) **do**
$G_r = G_r \cup Connect$ (S_i, S_j), where e is connecting the SSN of S_i and that of S_j;

end

Return (F, Gr);

End

4.3.3 Baseline Approaches

The performance of ORP is compared to the following three approaches. The first approach forms a minimum spanning tree based on a single-phase relay node placement [19] and we call it the Minimum Spanning Tree Approach (MSTA). The second is for solving a Steiner tree problem with minimum number of Steiner points [9]; we call it Steiner with Minimum Steiner Points (SwMSP). The third approach is the Grid-based Optimized Relay Placement (GORP) as described previously. The MSTA opts to establish an MST through RN placement. It first computes an MST for the given WSN partitions (SSNs) and then places RNs at the minimum number of grid vertices on the MST in accordance to Algorithm 4.1. The SwMSP approach pursues a Steiner tree model; it places the lowest relays count to maintain connectivity such that the transmission range of each node is at most r (i.e., the maximum edge length in the Steiner tree is $\leq r$). SwMSP first combines nodes that can directly reach each other into one Connected Group (CG). The algorithm then identifies for every three CGs, there is a vertex x on the grid that is at most r(m) away. An RN is placed at x and these three CGs are merged into one CG. These steps are repeated until no such x could be identified (i.e., no disconnected group). After that, each group is represented as a point y and an MST is computed based on the y points. Accordingly, the total number of populated relays using the SwMSP approach is

$$Q_{RN} = X + \left(\frac{L}{r} - 1\right) \qquad (4.12)$$

where X is the count of x points, and $\left(\dfrac{L}{r}-1\right)$ is the total relays popu-
lated on each edge of the computed MST (where L is the length of the edge). While the total number of populated relays using MSTA and ORP equals Q_{FPRNs} and $Q_{FPRNs} + Q_{SPRNs}$, respectively.

In summary, all MSTA, SwMSP, and GORP deployment approaches are used as a baseline in this research due to their efficiency in linking WSN partitions while maintaining the minimum number of relays required in the network federation.

4.3.4 Simulation Model

The four deployment schemes: MSTA, SwMSP, GORP, and ORP, are executed on 500 randomly generated WSN graph topologies in order to get statistically stable results. The average results hold confidence intervals of no more than 2% of the average values at a 95% confidence level. For each topology, we apply a random node/link failure based on pre-specified PNF and PLF values, and performance metrics are computed accordingly. A Linear Congruential random number generator is used. Dimensions of the deployment space vary from 50 to 250 (km²). We assume a predefined fixed time schedule for traffic generation at the deployed WSN nodes. Relays positions are found by applying the four deployment strategies. To simplify the presentation of results, all the transmission ranges of sensors and relays are assumed equal to 100 m.

4.3.5 Simulation Results

For a fixed number of disjoint sectors (=3) and deployment area (=50 km²), Figure 4.5 compares ORP approach with MSTA, SwMSP, and GORP in terms of the federated WSN sectors connectivity. It shows how ORP and GORP outperform the other two approaches under different PNF/PLF values. Unlike the other two approaches, WSNs federated using the GORP approach stay connected even under PNF = PLF = 50%. This is a very desirable behavior in harsh environments targeted by large-scale OEM applications. However, it shows a rapid decline in the network connectivity while the PNF/ PLF values increase, which is not the case with the ORP approach.

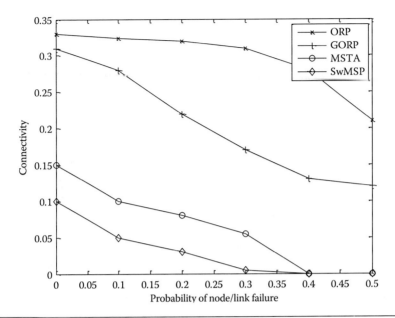

Figure 4.5 Connectivity vs. the PNF/PLF.

Applying ORP provides a noticeable steady state connectivity while node/link failure increases. This can be returned to the larger feasible search space that has been ignored while assuming the virtual grid. Moreover, connectivity levels achieved by the ORP outperform the levels achieved by the GORP due to considering the network connectivity while forming the network backbone in the first phase of the ORP approach.

Figure 4.6 depicts the effects of the RN count on the inter-connectivity of the federated WSN sectors. It shows the average λ_2 (i.e., connectivity) for the federated WSNs using different total numbers of RNs, where the number of disjoint sectors is fixed to 3 in order to see the effect of the relay node placement, and PLF = PNF = 0.2. It is clear how an increment in the deployed RNs leads to a rapid increment in connectivity even in the presence of 20% nonfunctional nodes/links using the GORP approach. Moreover, using 15RNs only, GORP achieves a connectivity value higher than the connectivity value achieved by the MSTA and SwMSP using 30 RNs; this indicates a greater saving in terms of the network cost. Nevertheless, more savings are reached while applying the ORP approach. This is

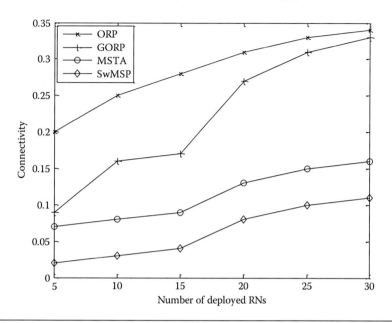

Figure 4.6   Connectivity vs. the QRN.

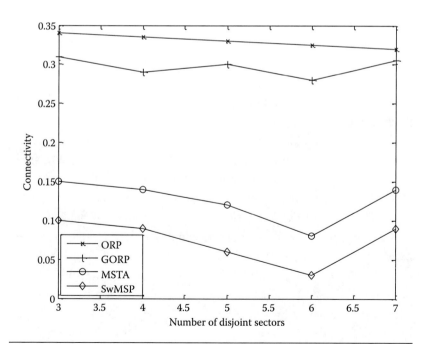

Figure 4.7   Connectivity vs. the QSSN.

also because of the consideration of connectivity since the early stages of the deployment (i.e., while constructing the backbone).

In Figure 4.7, GORP consistently outperforms MSTA and SwMSP with various disjoint sectors (i.e., different Q_{SSN} values) and large PNF and PLF (=40%). This is due to the placement of the SPRNs in GORP which always aims to maximize the federated sectors connectivity regardless of their counts. It is worth noting that as the Q_{SSN} gets larger, the performance of MSTA and SwMSP becomes worse with such a large deployment area. However, ORP is not only providing better connectivity levels, but also shows a steady state output even while considering different sectors count. This has a great effect on the federated network scalability. We excuse the increase of connectivity when MSTA and SwMSP are used to federate more than six sectors by the dense distribution of sectors within a fixed deployment area (=100 km²). For more elaboration upon the effects of the deployment area, consider Figure 4.8.

Again, in Figure 4.8, GORP constantly outperforms MSTA and SwMSP, with varying deployment areas and PNF and PLF values equal to 20%, as long as the deployment area is within a reasonable

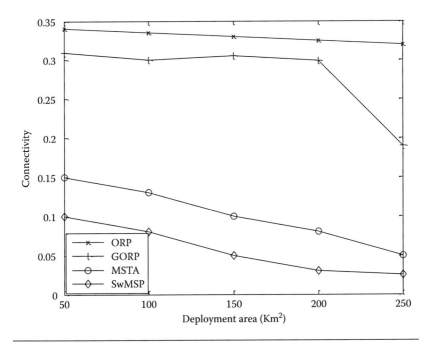

Figure 4.8   Connectivity vs. the DA.

size ($\leq$200 km^2). This gives more stability for the federated sectors in large-scale WSN applications. Even with a very huge area ($\geq$250 km^2), GORP is still much better than MSTA and SwMSP in terms of connectivity because of the deployed SPRNs. We observe that the sudden decrease in connectivity when we use the GORP approach is due to the lack of SPRNs with respect to the huge targeted area. However, this does not happen while applying ORP, which again outperforms GORP, because of the well planned FPRN deployment.

4.4 Conclusion

In this chapter, we look into the problem of deploying a fixed number of RNs to federate WSN sectors in OEM applications with the objective of maximizing network connectivity. An optimized two-phase approach is presented. The first phase utilizes some geometrical structures (namely, MST, DT, and Steiner tree) to construct a backbone of RNs that connect all WSN sectors, and finds a finite set of candidate locations for more RNs to be deployed in the second phase. The second phase deploys the remaining RNs in some of the candidate locations with the objective of maximizing connectivity of the network; this is done by solving a relaxed SDP.

The extensive simulation results, obtained under harsh operational conditions, demonstrate that the proposed two-phase strategy has the potential to provide tightly-connected networks that are suitable for environmental applications. Moreover, deployment strategies presented in this chapter can provide a tangible guide for network provisioning in large-scale environmental applications which require connecting vastly separated WSN sectors.

References

1. M. Younis and K. Akkaya, Strategies and techniques for node placement in wireless sensor networks: A survey, *Elsevier Ad Hoc Network Journal*, vol. 6, no. 4, pp. 621–655, 2008.
2. F. Al-Turjman, H. Hassanein, and M. Ibnkahla, Optimized node repositioning to federate wireless sensor networks in environmental applications, In *Proceedings of the IEEE Global Communications Conference (GLOBECOM)*, Houston, TX, 2011.

3. D. Hughes, P. Greenwood, G. Coulson, G. Blair, G. Pappenberger, F. Smith, and K. Beven, An intelligent and adaptable flood monitoring and warning system, In *Proceedings of the UK E-Science All Hands Meeting (AHM)*, Nottingham, UK, 2006.

4. M. Guizani, P. Müller, K. Fähnrich, A. V. Vasilakos, Y. Zhang, and J. Zhang, Connecting the world wirelessly, In *Proceedings of the ACM International Wireless Communications and Mobile Computing Conference (IWCMC)*, Leipzig, Germany, 2009.

5. N. Li and J. C. Hou, Improving connectivity of wireless ad hoc networks, In *Proceedings of the IEEE International Conference on Mobile and Ubiquitous Systems: Networking and Services (MobiQuitous)*, San Diego, CA, 2005, pp. 314–324.

6. A. Abbasi, U. Baroudi, M. Younis, and K. Akkaya, C2AM: An algorithm for application-aware movement-assisted recovery in wireless sensor and actor networks, In *Proceedings of the ACM International Wireless Communications and Mobile Computing Conference (IWCMC)*, Leipzig, Germany, 2009, pp. 655–659.

7. A. Bari, A. Jaekel, and S. Bandyopadhyay, Optimal placement of relay nodes in two-tiered, fault tolerant sensor networks, In *Proceedings of the IEEE International Conference on Computers and Communications (ICCC)*, Aveiro, Portugal, 2007, pp. 159–164.

8. E. L. Lloyd and G. Xue, Relay node placement in wireless sensor networks, *IEEE Transactions on Computers*, vol. 56, no. 1, pp. 134–138, 2007.

9. X. Cheng, D.-Z. Du, L. Wang, and B. Xu, Relay sensor placement in wireless sensor networks, *Wireless Networks*, vol. 14, no. 3, pp. 347–355, 2008.

10. K. Xu, Q. Wang, G. Takahara, and H. Hassanein, Relay node deployment strategies in heterogeneous wireless sensor networks, *IEEE Transactions on Mobile Computing*, vol. 9, no. 2, pp. 145–159, 2010.

11. K. Akkaya, M. Younis, and M. Bangad, Sink repositioning for enhanced performance in wireless sensor networks, *Computer Networks*, vol. 49, pp. 512–434, 2005.

12. A. Abbasi, M. Younis, and K. Akkaya, Movement-assisted connectivity restoration in wireless sensor and actor networks, *IEEE Transactions on Parallel Distributed Systems*, vol. 20, no. 9, pp. 1366–1379, 2009.

13. S. Lee and M. Younis, Optimized relay placement to federate segments in wireless sensor networks, *IEEE Transactions on Selected Areas in Communications*, vol. 28, no. 5, pp. 742–752, 2010.

14. F. Al-Turjman, A. Alfagih, H. Hassanein, and M. Ibnkahla, Deploying fault-tolerant grid-based wireless sensor networks for environmental applications, In *Proceedings of IEEE Conference on Local Computer Networks (LCN)*, Denver, CO, 2010, pp. 731–738.

15. J. O'Rourke. *Computational Geometry in C*, Cambridge University Press, Cambridge, UK, 1998.

16. H. J. Prömel and A. Steger. *The Steiner Tree Problem: A Tour through Graphs, Algorithms, and Complexity*, Vieweg +Teubner, Wiesbaden, Germany.
17. A. Ghosh and S. Boyd, Growing well-connected graphs, In *Proceedings of the IEEE Conference on Decision and Control*, San Diego, CA, 2006, pp. 6605–6611.
18. S. Boyd, Convex optimization of graph Laplacian eigenvalues, *Proceedings of the International Congress of Mathematicians*, vol. 3, no. 63, pp. 1311–1319, 2006.
19. http://sdpa.indsys.chuo-u.ac.jp/sdpa/archive.html.

5

TOWARDS AUGMENTING FEDERATED WIRELESS SENSOR NETWORKS IN FORESTRY APPLICATIONS*

Many factors are attributed to the wide use of Wireless Sensor Networks (WSNs) in today's applications and across different environments. Improvements in Micro-electromechanical Systems (MEMS), transceiver hardware, sensing platforms, and energy harvesting have all aided the design of more efficient WSNs. As such, WSNs are now deployed in many capacities over various domains, most notably in Environmental Monitoring (EM). Generally identified as harsh environments for WSNs, they extend to cover natural disasters such as volcanoes, floods [1], and forest fires [2].

As these scenarios encompass many harsh physical factors, they are more prone to failures. The scope of failure does not affect nodes in singularity, but it often affects significant sectors of the deployed WSN, causing sizable partitioning in the underlying topology. Since multiple networks are often deployed at large to serve multiple applications, it is imperative to maintain connectivity between them to achieve the global goal of efficient and real-time monitoring of that environment. This entails maintained connectivity even under high probabilities of failure. We refer to failures at the node level as PNF and at link levels as PLF. Understanding the performance of the global network under these failures and the resulting connectivity measures dictate the effectiveness of the WSN witnessing federation.

Establishing, or often re-establishing, connectivity has been approached in multiple ways in the literature of WSNs. Mainstream

---

* This chapter has been coauthored with Hossam Hassanein, Sharief Oteafy, and Waleed Alsalih.

approaches include deploying relay nodes (RNs) to establish (often multiple) paths in the network [1,3] as a whole, or utilizing mobile nodes that are able to reconnect partitions by moving into a median location [4,5]. A solution scenario of the former is seen in Reference [6] whereby a minimum threshold for relay nodes is established to deploy in a disconnected static WSN to regain connectivity. Another example of the latter [4] adopts mobile nodes that would relocate to establish *k*-connectivity properties, as required by the application. This is also optimized by determining a minimum on the number of nodes that need to re-position to establish this metric.

The previous approaches could be utilized in benevolent environments, where probabilities of failure are under control or where a pattern which we can model is seen. Nevertheless, in forestry EM scenarios, re-establishing connectivity between federated networks entails more hindrances. Dominantly, the irregularity of communication regions of such networks not adhering to the regularly assumed disc-shape [7] dictates a hard-to-model partitioning problem, hence making the establishment of reconnection zones a major issue. Challenges from significant distances between sectors, which might reach further than twice the communication range of an RN, added to the cost of RNs are among the dominant hindrances.

Our contribution solves the problem of RN placement by dissecting the problem to polynomial time operations to leverage the intractable issues with the inherently complex reconnection problem. We adopt a two-fold approach to this problem, namely Fixing Augmented network Damage Intelligently (FADI). The base protocol adopts a grid-based approach to dissect the search space into a set of finite points where RNs would be deployed to reestablish connectivity. The second fold uses the derived sets of points to determine the optimal assignment of RNs to these points, minimizing the number of relay nodes while maintaining connectivity. Moreover, the PNF and PLF metrics form constraints on the optimization problem.

The remainder of this chapter is organized as follows: Section 5.1 outlines the background to this problem and the related work carried out in reconnecting federated WSNs. Then, Section 5.2 details the system level assumptions and parameters, all invoked by the problem definition. The proposed approach (FADI) is presented in detail in Section 5.3 with elaborated explanations and full pseudocode for all the underlying algorithms in this approach. This is followed by a performance evaluation in comparison to two dominant contenders in this domain in Section 5.4. The chapter is concluded in Section 5.5 with directions of future work.

5.1 Background and Related Work

Failures in WSNs is a common phenomenon that requires low-cost and real-time maintenance schemes. One of the most common failures is loss of links which hinders network communication and can result in complete network partitioning. In networks where co-processing takes place, especially when information fusion is utilized [8], network partitioning can be detrimental to their operations. WSNs can be federated either by employing mobile nodes in the originally deployed network [9], or by populating a few relay nodes based on the network damage size. In this chapter, we focus on the latter approach due to its cost-efficiency and applicability in outdoor large-scale forestry environments. In Reference [10], Lloyd and Xue opt to deploy the fewest RNs such that each sensor node is connected to at least one RN. Additionally, the inter-RN network is strongly linked by forming a Minimum Spanning Tree (MST), and employing a Geometric Disk Cover algorithm. In Reference [11], the authors solve a Steiner tree problem to deploy the fewest RNs. Although the Steiner tree approach may guarantee the best network topology, it may not encompass the minimum number of relay nodes, as shown in Section 5.4.

Unlike [10,11], Xu et al. [12] study a random RN deployment that considers network connectivity for the longest WSN operational time. The authors proposed an efficient WSN deployment that maximizes network lifetime when RNs communicate directly with the Base Station (BS). In this study, it was established that different energy consumption rates at different distances from the BS render uniform RN deployment, a poor candidate for network lifetime extension. Alternatively, a weighted random deployment approach is proposed. In this random deployment, the density of RN deployment increases as the distance to the BS increases; thus, distant RNs can split their traffic amongst themselves. This in turn extends the average RN lifetime while maintaining a connected WSN.

In Reference [13], a distributed recovery algorithm is developed to address specific connectivity degree requirements. The contribution is identifying the minimal set of relay nodes that should be repositioned in order to reestablish a particular level of connectivity. Nevertheless, these references (i.e., [12,13]) do not minimize the relay count, which may not be cost effective in forestry applications.

In contrast, considering both connectivity and relay count was the goal of Reference [14]. In it, Lee and Younis focus on designing an optimized approach for federating disjointed WSN segments (sectors) by populating the least number of relays. The deployment area is modeled as a grid with equal-sized cells. The optimization problem is then mapped to selecting the fewest count of cells to populate relay nodes such that all sectors are connected.

Unlike [14], in this chapter, FADI considers the relay count and neighborhood degree in a different model. It derives positions of highest potential in establishing connectivity between the disjointed functional nodes. This, in, turn renders a more time efficient approach than those generated by References [10,11], and unlike [12,13], FADI addresses the network federation problem without violating WSN cost-effectiveness.

5.2 System Model

The system is inherently designed as an augmentation approach for connectivity reestablishment in forestry applications. As such, it is important to rigorously define the parameters of the system according to which our model operates. This section presents the problem definition in terms of how it would relate to a general forestry application and the framework of the required solution. An elaborate explanation follows to highlight the network parameters upon which our system operates, the governing communication metrics, and the foundation for the grid-based approach adopted in the model.

5.2.1 Problem Definition

Consider a WSN which underwent partitioning into multiple sectors, often the result of a physical phenomenon or cascading failure in a given region. Without loss of generality, we assume that each sector is represented by a node which we call a functional node. This is justi-fied by locating the nearest node to the border of the damaged region which is connected to the rest of its sector. As such, re-establishing connectivity with that node will reintroduce a path to every other node in that sector. Thus, our problem is formulated as follows:

Given n functional nodes, determine the minimum number of relay nodes required to establish intra-functional node connectivity, and their positions respective to the network at large.

An optimal solution to this problem can be derived from the Minimum Spanning Tree (MST) algorithm which is especially beneficial due to its dominant polynomial time solution (e.g., using Kruskal's algorithm). Forming an MST involves spanning the func-tional nodes with minimum weight edges where the weight reflects the number of relays required to establish an edge between two func-tional nodes. We note that edge weight in this research reflects the cost of establishing an edge between the disjointed functional nodes. In forestry applications, relay nodes are the most dominant expense in network hardware.

5.2.2 Communication Model

Evidently, establishing the metrics and bounds of communication are important to accurately resemble environmental monitoring applica-tions which we target in this research. Unlike traditional scenarios of flat earth communication and simplistic paradigms, many efforts have been made to identify the dominant factors of communication hindrance in long-range and outdoor treed environments like forests. It is certain that signal power faces significant decay as it travels for longer distances, yet this is difficult to model in our scenario domain. As per the constraints of our problem formulation, we have adopted the communication model presented in Reference [15], referred to as the log-normal shadowing model, since it accounts for irregular communication range scenarios. Thus, we represent the signal level at distance d from a given transmitter as:

$$P_r = K_0 - 10\gamma \, \log(d) - \mu d \tag{5.1}$$

which follows a log-normal distribution centered around the aver-age power value at that point. Here K_0 is a constant incurred at transmission (of transceiver electronics), which is derived from the mean heights of T_x and R_x. Having d as the Euclidean distance (in 3-D space) between the transmitter and receiver and γ as the path loss exponent, we adopt μ as a normally distributed random variable (r.v.) with zero mean and variance, i.e., $\mu \sim N(0, \sigma^2)$. Since the received signal could be quantified using P_r, we devise a lower threshold on the signal level to deem communication successful. Denoting it as P_{min} over distance d (between transmitter and receiver), we denote the probability of successful communication as:

$$P_c = P\left(P_r(d) \geq P_{min}\right) \qquad (5.2)$$

which could be presented, after substitution from Equation 5.1 and algebraic manipulations, as:

$$P_c(d, \mu) = K_0^{-d\mu} \qquad (5.3)$$

where $K_0 = 10\log(K)$. This equation emphasizes the important role of surrounding factors in the environment in signal depletion due to obstacles and terrain properties, not simply the direct relationship with distance. Thus, we formalize the connectivity (symmetric communication) between two nodes in the network as:

Definition 5.1: A "probabilistic connection" exists between two nodes, of distance d apart, if for a given threshold parameter τ we assert that $P_c(d, \mu) \geq \tau$, where $0 \leq \tau \leq 1$.

5.2.3 Network Model

We regard the network as a heterogeneous WSN with two tiers. The first is formed by functional nodes; the second is a layer of RNs which establish long-range communication across the network and to the sink. Nevertheless, where partitioning happens and when the topology dictates long range communications, optimal placement for RNs is adopted to federate the network. It is also important to note that the involvement of MAC protocols and the impact of network partitioning and federation on them are of great significance. However, recent advancements in adopting efficient MAC protocols for real time environments, as presented by Egea-López et al. [16], render this topic beyond the scope of our research.

Since network lifetime is an aggregation of that of its nodes, we elaborate on what terminates lifetime in this model. It is important to note that generally in forestry applications, PLF is quite elevated. As such, a fully operable node with significant energy reservoir may become useless to the network when its communication capabilities are jeopardized. Simply measuring the remaining pool of energy at nodes is not a significant indicator. Instead, the most applicable and realistic measure of lifetime would take into consideration the connectivity of its nodes. Hence, we formally define network lifetime as the duration before a partition occurs.

5.2.4 Grid Model

As the search space for possible locations for RNs is inherently intractable, the task of identifying the best candidate points is imperative yet not intuitive. Therefore, the protocol presented in this chapter identifies the potential positions for RNs as an initial phase in the solution. These positions are used in placing relays, which establish the MST based on a ranked approach, to achieve an optimal RN deployment scheme. Our approach here is adopting a 3-D grid which uniformly dissects the region covered by the network—into virtual cubes—and reduces the infinite search space to a discrete and finite set. Accordingly, the intersection points of these grid lines (cube corners) are referred to as grid unit centers. Eventually, RNs will only be positioned at a defined set of grid unit centers. As such, we describe the potential of a grid unit center as a candidate for RN placement according to the nodes able to communicate to that point. This is based on the communication metrics outlined in Section 3.2. Formally, we define the nodal coverage of a grid center as:

Definition 5.2: Given a grid unit center c in the deployment space, a functional node x is said to cover c if and only if $P_c(x, y) \geq \tau$. Where y is a RN placed at c.

Thus, we are able to quantify the potential of a grid center in terms of its connectivity, as an aggregation of the nodes covered by it; more formally:

Definition 5.3: Connectivity potential of a grid center c is proportional to the sum of functional nodes covering c.

Hence, choosing the most representative parameters for the communication model of the network, we are able to assign a set of potential nodes for each grid unit center which cover it and use that as a ranking scheme for optimal locations for RN redeployment. Formally:

> **Definition 5.4:** The Grid Unit Potential Set (GUPS) of a grid center c holds all the functional nodes covering c. The degree $D = |GUPS|$. It is Maximal GUPS (MGUPS) if there exists no set θ s.t. $|MGUPS| \subseteq \theta$. The subset of functional nodes' coordinates connected to c is denoted $S(c)$.

5.3 Fixing Augmented Network Damage Intelligently (FADI): The Approach

The proposed approach, namely Fixing Augmented network Damage Intelligently (FADI) is presented in this section. Given a network facing significant dissection/partitioning, we underline the procedural approach to efficiently locate positions for RN placement and detail the scheme for reconnecting it. Since the system is based on interchangeable operating procedures, the protocols are elaborately explained and their formal algorithms are presented. Figure 5.1 depicts our approach in light of the algorithms presented below; an

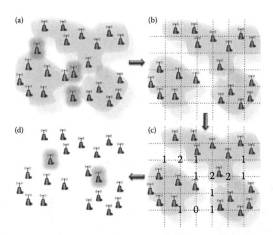

Figure 5.1   Depicts the operation of FADI. (a) In a WSN where some pivotal nodes are lost (with glowing background), federation is required to reestablish connectivity, (b) grid construction takes place to test grid unit points for node coverage, shown in (c), where the candidate MGUPS are chosen, resulting in the final deployment plan highlighted in (d) to restore minimum connectivity.

example of a WSN which has undergone partitioning and the federation steps is highlighted.

The first phase of the approach identifies the set of grid centers that
would satisfy the optimality highlighted earlier. Then the connectivity potential of each is determined. Following that, the MGUPs are
derived and the MST approach is utilized to federate the network.
Having the set of functional nodes to start with, we present the itera-
tive approach as proposed in Algorithm 5.1. Moreover, this meta-
algorithm invokes the procedures highlighted in Algorithms 5.2
through 5.5, serving all together to identify MGUPS and assigning
RNs to them until connectivity is re-established. The constraints,
inputs, and outputs are all detailed in their respecitve algorithms as
presented below.

Algorithm 5.1 provides the high level pseudo-code of FADI. The
algorithm takes the list of nodes which are still functional as an input.
The detailed pseudo-code of finding the grid centers' potential for
connectivity, the grid centers of the highest potential, and the set of
grid centers used to place the relays constructing the MST are shown
in Algorthims 5.2 through 5.5. Algorithm 5.2 associates each grid
center with its connectivity potential. In lines 9–14 of Algorithm 5.2,
FADI computes the probability of the grid unit center i being connected with each functional node individually based on Equation 5.3.
This is repeated by lines 7–15 until all probabilities between the grid
units' centers and all functional nodes are computed. Based on these
probabilities, set C is initialized in line 12. Algorithms 5.3 and 5.4
check for grid centers that have the highest potential for connectivity
(see Definition 5.3), which we call MGUPs.

Algorithm 5.4 calls Algorithm 5.3 to test whether the set of functional nodes covering a specific grid unit center is maximal or not. In
line 7 of Algorithm 5.3, FADI searches for any set (other than C_i) in
C that has the same functional nodes which cover the grid center i.
If such set is found, Algorithm 5.3 returns false. Otherwise it returns
true, meaning that set C_i is maximal. After discovering the grid centers which have the highest potential for connectivity (i.e., set M in
line 9 of Algorithm 5.1), FADI calls Algorithm 5.5 to construct the
MST using these grid centers (or MGUPs). Line 10 of Algorithm 5.5
carries out a search for the two closest functional nodes. If the two
closest nodes are not connected (i.e., $P_c \leq \tau$), it looks at line 12 for

the minumum number of grid centers in which the relays have to be placed to connect these two nodes. After connecting the two closest functional nodes (i.e., a connected component), we iteratively look for the next closest functional node that has to be connected to them. And in lines 20–36 of Algorithm 5.5, FADI iterativley searches for the least grid centers to be used for placing the relay nodes that will connect the next closest node with the connected component.

Algorithm 5.1: General operation of FADI

1. **Function FADI** (*F*: Initial Set of nodes to construct FN)
2. **Input:**
3. A set *F* of the functional nodes' coordinates.
4. **Output:**
5. A set *GC* of the intelligently selected
6. grid centers for RNs placement.
7. begin
8. **while** WSN is partitioned **do**
9. $C=$FindGridUnitsPotential(F);
10. $M=$**FindMGUPs**(F, C);
11. $GC=$**IMST** (F, M);
12. **endwhile**
13. **end**

Algorithm 5.2: Test grid unit potential for connectivity repair

1. **Function FindGridUnitsPotential**(F)
2. **Input:**
3. A set *F* of the functional nodes' coordinates.
4. **Output:**
5. A set $C=\{C_i \mid \forall \, i \in$ grid units' centers$\}$.
6. **Begin**
7. **foreach** grid unit center *i* **do**
8. $C_i := \varnothing$;   //the set of covered functional nodes by center *i*.
9. **foreach** functional node *j* **do**

10.   Compute $P_c(i, j)$;

11.   **If** $P_c(i, j) \geq \tau$

12. $C_i := j \cup C_i$;   //Add the coordinates of j to C_i

13.   **endif**

14.   **endfor**

15.   **endfor**

16.   **End**

Algorithm 5.3: Testing whether a grid unit set C_i is maximal or not

---

1.   **Function Maximal(C_i, C)**

2.   **Input:**

3.   A set C_i of functional nodes covering a specific grid

4.   unit center i and a set C of all non-empty C_i sets.

5.   **Output:**

6.   True if C_i is MGUPS and False otherwise.

7.   **Begin**

8.   Search for a set C' such that $C_i \subseteq C'$.

9.   **If** $C' := \varnothing$ **do**

10.   **return** True;

11.   **else**

12.   **return** False;

13.   **endif**

14.   **End**

Algorithm 5.4: Finding all Maximal Grid Unit Potential Set MGUPS

---

1.   **Function FindMGUPs(F, C)**

2.   **Input:**

3.   A set F of the functional nodes' coordinates.

4.   A set C of all non-empty C_i sets.

5.   **Output:**

6. A set M that contains one position from every MGUPS.

7. **Begin**

8. $M := \emptyset$;

9.   **foreach** $C_i \in \{C\}$ **do**

10.   **If Maximal**(C_i, C) **do**

11. $M := \{i\} \cup M$;

12.   **endif**

13.   **endfor**

14. **End**

Algorithm 5.5: Intelligent Minimum Spanning Tree (IMST)

---

1. **Function IMST** (F, M)

2. **Input:**

3. A set F of the functional nodes' coordinates.

4. A set M of the MGUPs.

5. **Output:**

6. A set GC of the intelligently selected grid centers for RNs placement.

7. **begin**

8. $GC := \emptyset$;

9. $FN := \emptyset$;

10. **If** $P_c(i, j) \leq \tau$

11. *Search for the least grid centers c required to connect i and j, such that* $c \in M$ **and** $P_c(i, c) \geq \tau$ **and** $P_c(c, j) \geq \tau$;

12. $GC := c \cup GC$; //Add the coordinates of j to GC//

13. $FN := c \cup FN$;

14. $F := F - I - j$;   //Set of Functional nodes//

15. **endif**

16. N=no. of remaining F nodes which are not in FN;

17. $i=0$;

18. **foreach** remaining node n_i in F **do**

19.   Find co_i;

20. $i=i+1$;

21.   **endfor**

22.   **Let $CO=\{co_i\}$**

23. $i=0$;

24.   **while** $i < N$ **do**

25. $S=$Smallest co_i in **CO**;

26. $GC := S \cup GC$;

27. $FN=FN \cup S \cup n_i$;

28. $F=F-n_i$;

29. $CO=CO-co_i$;

30. $i=i+1$;

31.   **foreach** remaining node n_j in F **do**

32.   Find co_j;

33. $j=j+1$;

34.   **endfor**

35.   **endwhile**

36.   **end**

To demonstrate the optimality of our approach, we introduce the following definition:

Definition 5.5: A finite set of positions P is optimal if there exists an ideal[*] placement of the least *RNs* in which each relay is placed at a position in P.

Accordingly, we have to show that the set GC in Algorithm 5.1 is ideal and derived from the set of MGUPS. Thus:

Lemma 5.1: For every GUPS β, there exists an MGUPS α such that $S(\beta) \subseteq S(\alpha)$.

Proof: If β is an MGUPS, we choose α to be β itself. If β is not an MGUPS then, by definition, there exists a GUPS α_1 such that $S(\beta) \subseteq S(\alpha_1)$. If α_1 is an MGUPS, we choose β to be

---

[*] Ideal in terms of connectivity degree.

α_1, and if α_1 is not an MGUPS then, by definition, there exists another GUPS α_2 such that $C(\alpha_1) \subseteq C(\alpha_2)$. This process continues until MGUPS α_x is found; we choose α to be α_x. Thus, Lemma 5.1 holds. ∎

Theorem 5.1: A set P that contains one position from every MGUPS is optimal.

Proof: To prove this Theorem, it is sufficient to show that for any arbitrary placement Z we can construct an equivalent[*] placement $\bar{Z}$ in which every RN is placed at a position in P. To do so, assume that in Z, an RN i is placed such that it is connected to a subset J of functional nodes. It is obvious that there exists a GUPS β, such that $J \subseteq C(\beta)$. From Lemma 5.1, there exists MGUPS α such that $C(\beta) \subseteq C(\alpha)$. In $\bar{Z}$, we place i at the position in P that belongs to α, so that i is placed at a position in P and is still connected with all functional nodes in J. By repeating for all minimum number of RNs, we construct a placement $\bar{Z}$ which is equivalent to Z, and thus Theorem 5.1 holds. ∎

Lemma 5.2: The GC set found by FADI is unique and has the least D.

Proof: This can be proven by contradiction. Assume FADI can find relay node placement A which contains the positions of the least RNs required to establish edges between the disjointed functional nodes. For contradiction, assume A is not unique. Then, there is another placement B in which the same relays' count is used. Let e_1 be an edge that is in A but not in B. As B forms an MST, $\{e_1\} \cup B$ must result in a cycle C in the federated network graph. Then B should include at least one edge e_2 that is not in A and lies on C. Assume the weight of e_1 is less than that of e_2. Replace e_2 with e_1 in B yields the spanning tree $\{e_1\} \cup B - \{e_2\}$ which has a smaller weight compared to B, thus a contradiction, as B was assumed to be a MST, yet it is not. ∎

---

[*] Equivalent in terms of connected functional nodes. In other words, the placement of an RN at position i, within the communication range of nodes x and y, is equivalent to the placement of the same RN at position j which is within the communication range of the nodes x, y, and z.

As for time complexity of the proposed approach, consider the following:

Lemma 5.3: Finding an MGUPS takes at most $(n - 1)$ step, where n is the functional nodes' count.

Proof: By referring to the proof of Lemma 5.1, it is clear that $|C(\alpha_x)| \leq n$, and $|C(\alpha)| \langle |C(\alpha_1)| \langle |C(\alpha_2)| \langle \cdots \langle |C(\alpha_x)| \leq n$; where $|C|$ is the cardinality of C. Consequently, the process of finding the MGUPS α_x takes a finite number of steps $\leq n-1$. ∎

Thus, the following Theorem holds:

Theorem 5.2: The run time complexity of the FADI approach is $O(n^2\log n)$, where n is the number of functional nodes.

Proof: Let g be the total number of grid unit centers on the assumed grid model, which is constant and known in advance for a specific monitored site. Since the total functional nodes is equal to n, the time complexity of Algorithm 5.2 is $O(gn)=O(n)$. in Algorithm 5.3 we search for C' such that $C_i \subseteq C'$ and this can be achieved in $O(\log(g))=O(g)$. According to Lemma 5.3, Algorithm 5.4 will be executed in $O(n)$. As for Algorithm 5.5, the time complexity would be $O(n \log n)$ due to the nested loop in line 26. As Algorithm 5.1 iterates over Algorithms 5.2 through 5.5 n times in the worst case of network damage where none of the functional nodes are connected, Algorithm 5.5 dominates the time complexity of the while loop in Algorithm 5.1 and thus, time complexity of Algorithm 5.1 is $n*O(n\log n)=O(n^2\log n)$. Thus, the FADI approach time complexity is $O(n^2\log n)$. ∎

5.4 Performance Evaluation

Using MATLAB®, we simulate randomly generated WSNs which have a graph topology and consist of varying number of partitioned functional nodes[*]. We simulate realistic communication channel characteristics taken from experimental measurements in a densely treed environment [7].

---

[*] Random in terms of functional nodes' count and positions.

5.4.1 Performance Metrics and Parameters

To evaluate our FADI approach, we tracked the following performance metrics:

- *Average RN degree (D)*: This is the number of functional nodes in the neighborhood of an RN. It reflects the federated network reliability under harsh forestry characteristics. It gives an indication for the federated WSN robustness where higher node degree yields stronger connectivity and enables better load balancing.
- *Average RN count (Q_{RN})*: This represents the cost-effectiveness of the deployment approach and the main objective targeted by our approach.
- *Recovering time (RT)*: This is the time required to federate the disjointed functional nodes and remove any partitioning.
- *Time to Partition (TtP)*: This is the time span before the network experiences a partition after being federated.

Three main parameters are used in the performance evaluation:

- *Number of functional nodes (Q_{FN})*: This represents the complexity of the addressed problem.
- *Node density (ND)*: This measures the federated network scalability in large-scale forestry applications.
- *Probability of Failure (PoF)*: This is the probability of physical damage for the deployed node and the probability of communication link failure due to bad channel conditions; it uniformly affects any of the network nodes/links. We chose this parameter as it reflects the harshness of the monitored forest.

5.4.2 Baseline Approaches

The performance of FADI is compared to the following two approaches. The first algorithm forms a minimum spanning tree without considering the intersections of the irregular communication range [10] and we call it Minimum Spanning Tree Approach (MSTA). The second is for solving a Steiner tree problem with the minimum number of Steiner points [11], and we call it Steiner with Minimum Steiner Points (SwMSP). The MSTA opts to establish an MST based on the

Euclidean distance separating two functional nodes, bearing in mind that the communication range is depending on the distance only. It first computes an MST for the given WSN partitions and then places RNs at the minimum number of grid vertices on the MST. The SwMSP approach places the least relay count to repair connectivity such that the maximum edge length in the Steiner tree is $\leq r$. SwMSP first combines functional nodes that can directly reach each other into one Connected Component (CC). The algorithm then identifies that for every three CCs, there is a vertex x on the grid that is at most r (m) away from the CC boundary nodes. An RN is placed at x and these three CCs are merged into one CC. These steps are repeated until no partitioning in the network is found. In summary, both MSTA and SwMSP deployment strategies are used as baseline approaches due to their efficiency in linking WSN partitions using the minimum relays count.

5.4.3 Simulation Setup and Results

The three deployment schemes, MSTA, SwMSP, and FADI, are executed on 500 randomly generated partitioned networks for statistically stable results. The average results hold confidence intervals of no more than 2% of the average values at a 95% confidence level. For each topology, we apply a random PoF, and performance metrics are computed accordingly. A Linear Congruential random number generator is employed. Dimensions of the deployment space are $900 \times 900 \times 300$ (m^3). Based on experimental measurements [7], we set our communication model variables as shown in Table 5.1.

And μ to be a random variable that follows a log-normal distribution function with mean 0 and variance of 10. We assume a predefined fixed time schedule for traffic generation at the deployed WSN nodes.

Table 5.1 Parameters of the Simulated WSNs

PARAMETER	VALUE
τ	70%
r	100 (m)
PoF	35%
g_i^{RN}	100 (byte/round)
g_i^{SN}	10 (byte/round)
γ	4.8

To simplify the presentation of results, all the transmission ranges of functional nodes and relays are assumed equal to 100 (m).

For varying number of disjointed functional nodes, Figure 5.2 compares the FADI approach with MSTA and SwMSP in terms of the total required relays. It shows how FADI outperforms the other approaches under different complexities of the targeted federation problem. Unlike the other approaches, the required relay count increases slightly when the FADI approach is utilized as the total partitioned nodes are increasing.

This indicates more savings in cost which is very desirable in harsh environments targeted by large-scale forestry applications. Figure 5.3 depicts the efficiency of MST-based approaches in terms of time complexity with respect to other approaches such as the SwMSP. It is clear how an increment in the disjointed functional nodes leads to an exponential increment in the time required for recovering (federation) when a Steiner tree approach is utilized. This has a great drawback on forestry applications which are most often time sensitive.

Figure 5.4 justifies the optimality of FADI in terms of finding the least relay positions federating all partitioned functional nodes. The average relay degree achieved by FADI is much better than the average degree reached by MSTA and SwMSP with various counts for the disjointed functional nodes (i.e., different QFN values). This in turn provides a robust network topology structure under harsh operational

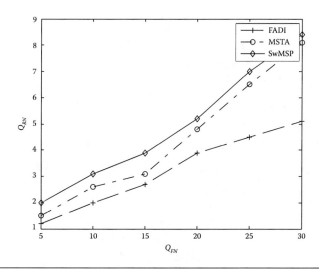

Figure 5.2   Functional node count vs. the required relays' count.

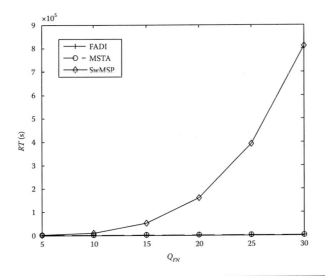

Figure 5.3 Functional nodes vs. the required recovery time.

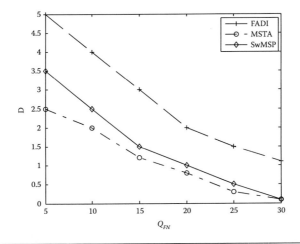

Figure 5.4 Functional nodes vs. the relays' neighborhood degree.

conditions in the forest. In Figure 5.5, we examined the practicality of our placement methodology under harsh operational conditions. FADI was able to generate a federated WSN that stays connected for long lifetime periods with respect to MSTA and SwMSP. It is also worth noting that FADI outperforms MSTA and SwMSP in terms of the relay degree, even in the presence of different node densities (ND) in the monitored site. This is shown in Figure 5.6; it points toward robust topologies in large-scale forestry applications where huge areas

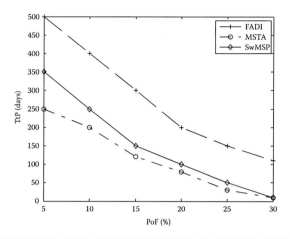

Figure 5.5   PoF vs. the time to partition.

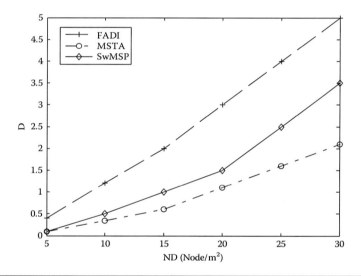

Figure 5.6   Node density vs. the relays' neighborhood degree.

are targeted with a relatively small number of sensor nodes (i.e., very small ND values).

5.5 Conclusion

In this chapter, we explored the problem of federating grid-based WSNs in forestry applications. A relay-based approach, called FADI, was presented using the minimum spanning tree algorithm. For practical solutions, varying probabilities of failures were considered in addition to limiting the huge search space of the targeted deployment problem. The extensive simulation results, obtained under harsh operational conditions, indicated that the proposed approach can efficiently federate the disjointed WSNs. Moreover, the deployment approach presented in this chapter can provide a tangible guide for network provisioning in large-scale environmental applications, which require linking between vastly separated WSN sectors. Future work would investigate the deployment problem in further environment monitoring scenarios. A subset of the relay nodes may have the mobility feature to repair connectivity and prolong the network lifetime. In addition, this work could extend to encapsulate a more inclusive framework for discovering nearby devices that may offer higher connectivity or act as intermittent relays, aiding in the federation scheme. The work presented by Gellersen et al. [17] facilitates a dynamic framework for discovering spatial relationships of devices with nearby (heterogeneous) devices.

References

1. D. Hughes, P. Greenwood, G. Coulson, G. Blair, G. Pappenberger, F. Smith, and K. Beven, An intelligent and adaptable flood monitoring and warning system, In *Proceedings of the UK E-Science All Hands Meeting (AHM'06)*, Nottingham, UK, 2006.
2. M. Younis and K. Akkaya, Strategies and techniques for node placement in wireless sensor networks: A survey, *Elsevier Ad Hoc Network Journal*, vol. 6, no. 4, pp. 621–655, 2008.
3. F. Al-Turjman, H. Hassanein, W. Alsalih, and M. Ibnkahla, Optimized relay placement for wireless sensor networks federation in environmental applications, *Wiley: Wireless Communication and Mobile Computing Journal*, vol. 11, no. 12, pp. 1677–1688, 2011.
4. A. Abbasi, U. Baroudi, M. Younis, and K. Akkaya, C2AM: An algorithm for application-aware movement-assisted recovery in wireless sensor and actor networks, In *Proceedings of the ACM International Wireless Communications and Mobile Computing Conference (IWCMC)*, Leipzig, Germany, 2009, pp. 655–659.
5. W. Alsalih, H. Hassanein, and S. Akl, Routing to a mobile data collector on a predefined trajectory, In *Proceedings of IEEE International Conference on Communications (ICC)*, Dresden, Germany, 2009, pp. 1–5.
6. N. Li and J. Hou, Improving connectivity of wireless ad hoc networks, In *Proceedings of the IEEE International Conference on Mobile and Ubiquitous Systems: Networking and Services (MobiQuitous)*, San Diego, CA, 2005, pp. 314–324.

7. F. Al-Turjman, H. Hassanein, and M. Ibnkahla, Connectivity optimization for wireless sensor networks applied to forest monitoring, In *Proceedings of the IEEE International Conference on Communications (ICC)*, Dresden, Germany, 2009, pp. AHSN11.5.1-5.

8. Y. Kim, S. Jeong, D. Kim, and T. S. Lopez, An efficient scheme of target classification and information fusion in wireless sensor networks, *Journal of Personal and Ubiquitous Computing*, vol. 13, no. 7, pp. 499–508, 2009.

9. F. Al-Turjman, H. Hassanein, and M. Ibnkahla, Optimized node repositioning to federate wireless sensor networks in environmental applications, In *Proceedings of the IEEE Global Communications Conference (GLOBECOM)*, Houston, TX, 2011.

10. E. Lloyd and G. Xue, Relay node placement in wireless sensor networks, *IEEE Transactions on Computers*, vol. 56, no. 1, pp. 134–138, 2007.

11. X. Cheng, D. Du, L. Wang, and B. Xu, Relay sensor placement in wireless sensor networks, *Wireless Networks*, vol. 14, no. 3, pp. 347–355, 2008.

12. K. Xu, H. Hassanein, G. Takahara, and Q. Wang, Relay node deployment strategies in heterogeneous wireless sensor networks, *IEEE Transactions on Mobile Computing*, vol. 9, no. 2, pp. 145–159, 2010.

13. A. Abbasi, M. Younis, and K. Akkaya, Movement-assisted connectivity restoration in wireless sensor and actor networks, *IEEE Transactions on Parallel Distributed Systems*, vol. 20, no. 9, pp. 1366–1379, 2009.

14. S. Lee and M. Younis, Optimized relay placement to federate segments in wireless sensor networks, *IEEE Transactions on Selected Areas in Communications*, vol. 28, no. 5, pp. 742–752, 2010.

15. T. Rappaport, *Wireless Communications: Principles and Practice*, 2nd ed., Prentice Hall, Upper Saddle River, NJ, 2002.

16. E. Egea-López, J. Vales-Alonso, A. Martínez-Sala, J. García-Haro, P. Pavón-Mariño, and M. Bueno Delgado, A wireless sensor networks MAC protocol for real-time applications, *Journal of Personal and Ubiquitous Computing*, vol. 12, no. 2, pp. 111–122, 2006.

17. H. Gellersen, C. Fischer, D. Guinard, R. Gostner, G. Kortuem, C. Kray, E. Rukzio, and S. Streng, Supporting device discovery and spontaneous interaction with spatial references, *Journal of Personal and Ubiquitous Computing*, vol. 13, no. 4, pp. 255–264, 2009.

6

OPTIMIZED HEXAGON-BASED DEPLOYMENT FOR LARGE-SCALE UBIQUITOUS SENSOR NETWORKS

Ubiquitous Sensor Networks (USNs) are a class of context-aware communication networks that provide an infrastructure for knowledge-based intelligent information service to anyone, anywhere, and at any time [1,2]. Wireless sensor networks (WSNs) form a basic and integral part of the USN infrastructure that provide sensed information to end users in diverse application environments such as agricultural monitoring in rural areas, structural health monitoring of buildings and bridges in urban areas, tracking items in industrial supply chain management applications, detection of forest fires, and even landmine detection in former war zones [3,4]. These USN applications require a large-scale deployment of the sensor network in order to cover the large target areas and provide more sensing points in the region being monitored. In such large-scale deployments, the network topology changes dynamically due to node deaths, changing node associations, and varying environment conditions, thus affecting the network connectivity and information gathering and delivery capabilities. In addition, the network may have to deal with service requests coming from a variety of end users, including individual consumers, public enterprises, government organizations, and even machines that monitor information. It is very challenging for current sensor networks to provide a common platform that can support such diverse USN applications, while providing context-aware information to end users that differ in their requirements on the attributes associated with service data such as reliability, latency, and throughput. As in the case with Internet of Things (IoT) applications [5,6], WSNs are not equipped to handle the heterogeneous traffic. Nor do they have adequate capacity to store the large volume of data generated as a result of the multiple requests being serviced by the network. They need modifications in the infrastructure to support the functionality [7–9]. Cloud computing offers a cost-efficient solution for USN data storage. It could serve as a single sink in the network that stores all the data generated in the network for further processing, or distributed cloud architecture could be used to serve as multiple data collection points in the network to help with distributed data processing [10,11].

To improve the capabilities of the network that delivers data to such a cloud, we propose the use of Cognitive Nodes (CNs) based on the elements of learning, reasoning, and knowledge representation for USN applications. CNs will provide enhanced capabilities to the WSN to not only deal with the network connectivity and node dynamics in large-scale deployments but also to provide local storage before reaching the final destination on the cloud. And this provides context- and user-aware service/data. To this end, the main contributions of this chapter are as listed below:

- We identify a grid-based deployment strategy for relay and cognitive nodes in the large-scale WSN such that the probability of successful data reception between the communicating nodes is >0.8.
- The use of a virtual grid in this study was mainly motivated by the ability to limit the huge search space of the network nodes' positions in large-scale applications such as those found in the IoT and ubiquitous sensing paradigms where everything shall be connected everywhere and at any time.
- We also calculate the number of relay and cognitive nodes required to cover the target area while ensuring a high probability of successful data reception.

The remaining sections have been organized as follows: In Section 6.1, we present the related work. Section 6.2 provides the details of the cognitive node architecture and details the system models. Then, we propose the deployment strategy for cognitive nodes in the network in Section 6.3. In Section 6.4, a case study is investigated to verify the proposed deployment strategy. Simulation results are presented in Section 6.5 before concluding the chapter in Section 6.6.

6.1 Related Work

Sensor node deployment problem has been extensively studied in the literature over the past decade. Researchers have considered various factors such as coverage, connectivity, energy efficiency, and fault-tolerance while proposing deployment strategies for sensor nodes (SNs) [12–17]. With the introduction of the ZigBee standard [18], the focus shifted from sensor node to relay node (RN) placement problem, as the RNs could serve to maintain connectivity of sensor nodes with their base station even when the network size scaled up [19–23]. The RNs increased the communication range of SNs and also took over the energy demanding task of data communication within the network from the SNs. This in turn increased the lifetime of the SNs, thus improving the longevity of the network. However, as WSN applications evolved from simple event monitoring or tracking applications to complex applications, such as monitoring a coal mine [24], network deployment and its operational complexity increased. The WSN had to not only provide periodically monitored data but also had to respond to on demand queries and emergency situations. The changing application requirements made the network traffic very heterogeneous, leading to load balancing issues among the nodes and traffic bottlenecks in the network. Recent research has even considered the use of mobile data collectors [25,26], traffic-aware relay node deployment, and artificial intelligence (AI) techniques to manage the dynamic network [27]. But data latency and reliability become an issue when mobile data collectors are used [28], and AI techniques have targeted very specific applications [29,30]. They have not been architecturally developed and implemented in a way that can be extended to different WSN application platforms. Thus, we say that in their current state, WSNs with SNs, RNs, and Data Collector nodes will not be able to understand and respond to changing application requirements. The network will not be able to cater to performance attributes of latency, reliability, energy consumption, and fault-tolerance while delivering data to the sink. We collectively call these attributes Quality of Information (QoI) attributes [31,32]; they represent the attributes that the application layer would associate with the data delivered to the sink, to measure the application-awareness of the response generated by the network to the end user's request. In order to make the network aware of the changing application requirements and enable it to provide QoI aware data, we propose the use of special nodes called Cognitive Nodes (CNs) in the underlying WSN. These CNs when strategically deployed in the network will ensure data delivery with user desired QoI to the sink in each round of data transmission throughout the lifetime of the network. We will refer to this network as an Information-Centric Sensor Network (ICSN) from this point forward. It will draw on the features of Information-Centric Networks (ICNs) in terms of named data association, in network caching, and the use of CNs as intermediate nodes that will process and store the information within the network [33,34]. However, we must mention that the idea of named data association in WSNs is not new. This idea has existed in Data-centric Sensor Networks (DCSNs) which are a special class of WSNs that function as information retrieval networks, rather than serving as point-to-point communication networks [3,35–39]. Sensor attributes are used for data gathering and delivery which makes the use of node addresses nonessential. This can lead to huge energy savings for the sensor network; a single query can be broadcast throughout the network to gather all relevant data from different sources vs. multiple queries addressed to specific locations to gather the same data. This translates to energy savings for all the network nodes, leading to prolonged network lifetime.

Shifting our focus back to the cognitive nodes, the information-centric approach to query dissemination used by these nodes helps find only relevant data and changes the way the network handles user requests. There is awareness in the network about the specific information requested by the user; i.e., temperature data or humidity information from a specific geographic area, at a specific time in the present, or from sometime in the past. The CNs enable the network to understand the QoI with which it is expected to return the requested data to the sink. In addition, the network is able to adapt the use of its resources to find paths that are either reliable, have low latency, or offer a high throughput. In this way, the network is not always exerting itself to find the best path that satisfies all the attributes. Rather, it prioritizes the QoI attributes for each transmission round based on the end user requirements and finds a suitable path accordingly, thus prolonging the network lifetime. Now, the challenge is in finding the best place to deploy these nodes in the ICSN. Optimal node placement is a very challenging problem and has been proven to be NP-hard [12]. With CNs, there are constraints on how many such nodes can be used in the network and whether one can only use CNs or combine it with the use of RNs in the underlying network.

In this chapter, we identify the cognitive functions of the CN, address its deployment problem, and through simulations, identify the best combination of RNs and CNs that the network can benefit from to minimize energy consumption and prolong network lifetime. All this is done while catering to the QoI attributes of reliability and instantaneous throughput. This contributes not only to the good quality of the user experience but also improves the lifetime of the network, during which data is delivered to the end user based on user desired QoI attributes.

6.2 System Models

6.2.1 Network Model

Nodes in WSNs can be deployed in a flat, hierarchical, or geographic location-based strategy. In terms of energy conservation, hierarchical deployment strategies provide better performance for WSNs [3]. We had proposed a hierarchical strategy for cognitive communication in WSNs in our earlier work [40,41]. We make use of a similar approach here for a large-scale WSN in USN applications. Cognitive Nodes (CNs), Relay Nodes (RNs), Sensor Nodes (SNs) and a sink node are the node level entities of the network. CNs act as cluster heads for RNs and are the decision makers for the network. They process the requests received from the end user by making use of the cognitive elements of knowledge representation, learning, and reasoning to identify a data delivery path to the sink that meets the user's QoI requirements [42]. RNs act as cluster heads for SNs and also participate in relaying information from CNs to the sink. SNs gather sensed data and forward it to both RNs and CNs lying within their communication range. The hierarchical deployment strategy helps to distribute the tasks between the RNs and CNs and better manage the network connectivity. We assume that RNs and CNs have the same communication range for a given transmit power. However, the transmit power for RNs is fixed (0 dB) at the time of deployment; CNs are allowed to adapt their transmission power from a pre-determined set of values (−3 to +10 dB) to achieve the desired transmission range and QoI.

6.2.2 Energy Consumption Model

The voltage discharge characteristics of most Lithium AA batteries (irrespective of their chemistry) suggest that once the terminal voltage drops to about 30% of its original value, almost all of the battery's usable energy is depleted [43]. Lithium batteries typically last for 500–1000 cycles before the terminal voltage drops to this value, depending on the application and environment in which it is operated [43]. In our system, we assume that the batteries at RNs and CNs are capable of delivering consistent performance for about 500 cycles, after which they are assumed to be drained out of energy. We set the initial battery cycle life to 500 units; every time a node is involved in a data or control message communication, we reduce the node's battery cycle life as shown in Table 6.1, based on the transmit power used for communication.

6.2.3 Communication Model

In this work, we use the ZunPhy transition region-based communication model [44] modified for outdoor environments. Overstepping the binary disc shaped model, the transitional region model identifies a location between the connected and disconnected regions in which the probability of having the received signal strength above a threshold value is above 80%. We use the log normal shadowing path loss communication model, with values for path loss exponent $n = 4$ and standard deviation of the zero-mean Gaussian random variable

Table 6.1   Reduction in Cycle Life Based on Transmit Power

P_t (dBm)	CYCLE LIFE REDUCTION (UNITS)
<3	1
3–5	2
5–7	3

$X\sigma$, $\sigma = 4$. The radios communicate in the ISM band at a data rate of 250 kbps and the reference distance $d0$ is 100 m. If d represents the distance between the transmitter and receiver, then the path loss (PL) at distance d is given by:

$$PL(d) = PL(d0) + 10n \log\left(\frac{d}{d0}\right) + X_\sigma \qquad (6.1)$$

where,

$$PL(d0) = 20\log\left(4\pi d0/\lambda\right) \qquad (6.2)$$

The received signal strength P_{recv} at distance d, for transmit power Pt is given by:

$$P_{recv}(d) = Pt - PL(d) \qquad (6.3)$$

Also,

$$\overline{PL(d)} = PL(d0) + 10\,n\log\left(\frac{d}{d0}\right) \qquad (6.4)$$

$$\overline{P_{recv}(d)} = Pt - \overline{PL(d)} \qquad (6.5)$$

6.2.4 Cost Model

In this section, we take a look at the cost of deploying relay and cognitive nodes in the WSN. The cost of a node is defined to be the total sum of the hardware cost and operational cost [45]. Operational cost includes the energy expended during the network operation and the delay involved in processing and transmitting the information from its source to the end user.

6.2.4.1 Relay Node

For a regular RN, this cost would be dominated by the cost of the node's radio (hardware), the energy consumed during data gathering and transmission, and the delay incurred in transmitting the information from its source to the end user. Thus, the cost of a relay node (C_{RN}) can be represented as follows:

$$C_{RN} = C_{radio} + C\left(RE_{Rx} + TE_{Tx}\right) + C(\delta_{trans}) \qquad (6.6)$$

C_{radio} represents the cost of the radio transceiver, which dominates the hardware cost of the RN. $C\left(RE_{Rx} + TE_{Tx}\right)$ represents the cost of data transmission and reception in terms of the energy expended by the node, where R and T are the number of packets that are received and transmitted during each transmission round and E_{Tx} and E_{Rx} represent the energy consumed during transmission and reception of the data packets. $C(\delta_{trans})$ represents the cost of delay incurred at the RN due to the hop-over-hop transmission time. If we represent the hardware cost as C_{RN-HW}, the energy cost as C_{RN-E} and the delay cost as $C_{RN-delay}$, the total cost of the RN C_{RN} can be written as follows:

$$C_{RN} = C_{RN-HW} + C_{RN-E} + C_{RN-delay} \qquad (6.7)$$

In the next subsection, we take a look at the cost incurred in deploying cognitive nodes in a WSN.

6.2.4.2 Cognitive Node   For a CN, the hardware, energy, and delay cost contribute to its total cost and we compare it with the cost of the RNs.

6.2.4.2.1 Hardware Cost (C_{CN-HW})   Cognitive nodes incur additional hardware cost in terms of their memory requirements (caching of sensor data), processor functions associated with learning and reasoning elements of cognition, and advanced radio equipment used for achieving increased transmit power and receiver sensitivity to improve the communication range and reliability. The increased communication range helps to reduce the total number of cognitive nodes used in the network and keep a check on the deployment cost.

$$C_{CN-HW} = C_{radio-TRX} + C_{cache} + C_{cog-proc.} \qquad (6.8)$$

Where, C_{CN-HW} represents the hardware cost per CN, C_{cache} is the cost of the cache memory per CN, $C_{cog-proc.}$ is the additional cost to the processor due to the elements of the cognitive process; i.e. learning and reasoning. $C_{radio-TRX}$ is the cost of using radio with improved receiver sensitivity and higher transmit power compared with the RN. Equation 6.8 can be expressed in terms of the hardware cost of RNs as follows:

$$C_{CN-HW} > C_{RN-HW} + C_{cache} + C_{cog.proc.} \qquad (6.9)$$

6.2.4.2.2 Energy Cost (C_{CN-E})   Typical functions of CNs that consume additional energy compared to regular RNs are data aggregation and the cognitive decision process. Additional energy consumption is accounted for by two factors: (1) protocol overhead incurred during cognitive data delivery due to feedback from the network during the learning process and the exchange of values of QoI attributes such as latency, reliability, and throughput while making routing decisions and (2) increased transmit power for increasing the communication range of CNs [46,47].

$$C_{CN-E} = C\left(TE_{Tx} + RE_{Rx}\right) + C\left(AE_{ag}\right) + C(PE_{cog.\,proc.}) \quad (6.10)$$

In Equation 10, *T, R, A,* and *P* are the total number of packets that are transmitted, received, aggregated, and processed by the cognitive elements respectively, in each transmission round. $C(TE_{Tx} + RE_{Rx})$ is the energy cost incurred during data transmission and reception, $C(AE_{ag})$ represents the energy cost incurred during data aggregation and $C(PE_{cog.\,proc.})$ represents the energy cost due to protocol and processing overhead during the cognitive processes. Expressing Equation 6.10 in terms of the energy cost of RNs we get:

$$C_{CN-E} \geq C_{RN-E} + AE_{ag} + CE_{cog.\,proc.} \quad (6.11)$$

If the relay and cognitive nodes use the same transmit power, then the equality sign holds true in Equation 6.11. In any case, the energy cost of the cognitive node is higher than that of the relay node.

6.2.4.2.3 Delay Cost ($C_{CN-delay}$)   Delay impacts the quality of the end user's experience. Hence, it is important to manage the network and node functions in such a way that delay is minimized and the end user's quality of experience is improved. Apart from the transmission latency, cognitive nodes incur additional delay in looking up the knowledge base for making cognitive decisions. Depending on the complexity of learning algorithm, the time taken to converge to a solution will add to the delay of the cognitive process. If $C_{CN-delay}$ is the delay per CN per transmission round, it can be represented as follows:

$$C_{CN-delay} = C(\delta_{trans}) + C(\delta_{lookup}) + C\left(\delta_{cog.\,decision}\right) \quad (6.12)$$

$C(\delta_{trans})$ is the cost of transmission delay and $C(\delta_{lookup})$ is the cost of delay incurred due to memory or cache look-up for matching sensor data and $C(\delta_{cog.decision})$ is the cost of the delay incurred during the cognitive decision making process. Equation 6.12 when expressed in terms of the delay at RNs can be written as:

$$C_{CN-delay} \geq C_{RN-delay} + \delta_{lookup} + \delta_{cog.decision} \qquad (6.13)$$

Thus, the total cost per cognitive node (C_{CN}) deployed in the WSN is given by:

$$C_{CN} = C_{CN-HW} + C_{CN-E} + C_{CN-delay} \qquad (6.14)$$

Using Equations 6.9, 6.11, and 6.13, Equation 6.14 can be expressed in terms of the cost of the relay node as follows:

$$C_{CN} > C_{RN-HW} + C_{RN-E} + C_{RN-delay} + \Delta \qquad (6.15)$$

This can be simplified and rewritten as:

$$C_{CN} > C_{RN} \qquad (6.16)$$

This equation reflects the additional cost of the cognitive nodes when compared with relay nodes. Hence, it is important to consider minimizing the number of cognitive nodes used in the deployment strategy in order to reduce the cost of network deployment and maintenance thereafter.

6.2.5 Problem Definition

For large-scale USNs, we define the node deployment problem as follows: Determine the number and location for the placement of relay and cognitive nodes in a given target area such that (1) the probability of the received signal strength is above a threshold value (−101 dBm) is 0.8 or more, (2) the network is connected in such a way that there is a path from each SN to the sink through the RNs or CNs at the time of deployment, and (3) high throughput and reliable data transmission over each hop till data is delivered to the sink. The probability of the received signal strength being above a threshold value is defined as the probability of successful data reception (P_r). It is a function of the separation distance d between two communicating nodes. When the separation distance is correctly estimated, the probability that the signal strength is above a specified threshold γ_{th} (the receiver's sensitivity for instance) can be estimated using a Q-function (Equation 6.17) based on the work in Reference [48].

$$Q(z) = \frac{1}{\sqrt{2\pi}} \int_z^\infty \exp\left(\frac{-x^2}{2}\right) dx \qquad (6.17)$$

Using the Q-function, the value of P_r can be estimated using a cumulative density function as follows:

$$P_r\left[P_{recv}(d) > \gamma_{th}\right] = Q\left[\frac{\gamma_{th} - \overline{P_{recv}(d)}}{\sigma}\right] \qquad (6.18)$$

As identified in the cost model, we want to minimize the number of CNs and keep their number lower than the number of RNs to minimize the total cost to the network. We also want to ensure that there is at least one RN/CN for each SN to deliver its information so that SNs are only involved in short-range, local communications that incur minimum cost. In this way, the network lifetime depends only on the RNs and CNs. In the following section, we identify a strategy for the deployment of CNs and RNs for large-scale USN applications.

6.3 The O2D Deployment Strategy

The node placement problem proposed in this chapter has an infinitely large search space and finding the optimal solution is an NP-complete problem. Therefore, we propose a 2-D hexagon grid model that limits the search space to a more manageable size [49]. We approximate the target area to be a hexagon region and divide the entire area into smaller hexagons of equal side L. We assume knowledge of the 2-D terrain of the targeted site ahead of time. Hence, the v candidate positions for RNs (on the hexagon vertices) are predetermined; unfeasible positions are excluded from the search space. We want to choose the optimum count and positions for *relay nodes* (RNs) among these v vertices given a specific number of SNs. Where these SNs are assumed to be uniformly distributed throughout the targeted area with a fixed communication range and transmission power, RNs have a fixed transmit power and communication range, but the values are higher when compared with those of SNs. The sink is assumed to be at the center of the target region. Given these assumptions, a hexagon grid-based deployment strategy, called *optimized 2-D deployment (O2D)*, is proposed in this section. The goal of the deployment strategy is to identify the length L of the side of each hexagon grid cell and the position of the RNs and CNs on the grid such that the RNs can communicate with at least one CN, and $P_r > 0.8$ along each hop of the data delivery path from a source node to the sink. To devise such a node deployment strategy, we make use of the following Lemmas and Theorem.

Lemma 6.1: Center of a hexagon is equidistant from each of its vertices as depicted in Figure 6.1.

Proof: Consider a hexagon with equal sides of length L. Diagonals of the hexagon have the property that they are of the same length and intersect at the center of the hexagon. This means that the half length of the diagonal from the center of the hexagon to any of its vertices is equal. This can also be proven from the congruence of the six smaller triangles formed by the intersecting diagonals, with RNs at the corners and the CN at the center, as seen in Figure 6.2. Thus, Lemma 6.1 holds. ∎

Lemma 6.2: Longest length of line from the center of a hexagon to any other point on the hexagon is given by the line segment joining the center with the vertices of the hexagon.

Proof: Consider a hexagon with equal sides of length L. A diagonal would divide the hexagon into six congruent triangles where the diagonal forms their twice side length. If we draw a circle with radius x, whose center lies at the center of the hexagon, we would be circumscribing a circle that passes through each of

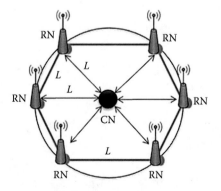

Figure 6.1   Hexagon cell of equal sides L having relay nodes (RNs) at the corners and a cognitive node (CN) at the center.

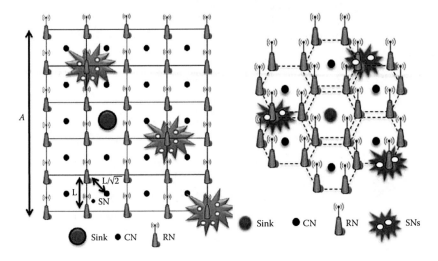

Figure 6.2 RNs and CNs on both square and hexagon grids.

the vertices of the hexagon. This circle does not touch any other point on any part of the hexagon except the vertices. Since the diagonal is the diameter of the circle, the radius is the maximum half length. And thus, $x = L$ and Lemma 6.2 holds. ■

Lemma 6.3: Received signal strength P_{recv} at a node increases as the distance between the communicating nodes decreases.

Proof: Let x be the distance between two communicating nodes. Using Equation 6.3, we say that $P_{recv}(x)$ represents the received signal strength at distance x. If the distance between the communicating nodes is doubled to $2x$, then the signal strength at the receiving node is represented as $P_{recv}(2x)$. For the same transmit power P_t and path loss at reference distance $PL(d0)$, the difference in the received signal strengths at distances $2x$ and x is approximated as follows:

$$P_{recv}(2x) - P_{recv}(x) = 10 * n * \log(x/2x) \tag{6.19}$$

Expressing $P_{recv}(x)$ in terms of $P_{recv}(2x)$, we get

$$P_{recv}(x) = P_{recv}(2x) + 3 * n \tag{6.20}$$

Thus, the received signal strength P_{recv} at distance x is increased by an amount of $3*n$, when compared with its value at double the distance $2x$. Thus, Lemma 6.3 holds. ■

Theorem 6.1: For a 2-D hexagon grid with nodes placed at the vertices, only one additional node placed at the center of the hexagon is sufficient to improve the probability of reliable reception of data between the vertices across the diagonal for the same transmit power.

Proof: For a hexagon with side of length L, from Lemma 6.1, we know that center of the hexagon is equidistant from the vertices. Using Lemma 6.1 and a corollary to Lemma 6.3, we can say that for a node placed at the center of the hexagon, the value of P_{recv} is approximately the same for data received from any of its corners. From Equation 6.18, we know that the probability of reliable data reception P_r is a function of P_{recv}. Thus P_r improves as P_{recv} increases for a given transmit power. Now, let us assume that there exists another point in the 2-D space around the hexagon, apart from the center, where a second node could be placed such that it is equidistant to all the vertices. This means that a second circle could be drawn with this point at the center such that it touches the hexagon only at its vertices. Then the distance between the diagonally opposite vertices would be the diameter of the circle, and its radius should be the longest distance between the vertex and this point. But we know that the diagonal of the hexagon is the longest length of line connecting opposite vertices. This means that the diameter is the same as the length of the diagonal. From Lemma 6.2, it follows that the half length of the diagonal is the radius of this circle and the center of this circle is the same as the center of the hexagon. Thus, we prove that only one additional node, lying at the center of the hexagon is sufficient to equally impact the probability of reliable data reception at each of the vertices. Thus, Theorem 6.1 holds. ∎

Theorem 6.2: The maximum number of vertices/corners in a grid of n hexagons in this study is given by $4n + 2$.

Proof: This Theorem can be proven simply by induction, where the first hexagon in the center of the grid contributes 6 vertices and each new hexagon next to it contributes at most 4 vertices. And thus a grid of n equal hexagons will have at most $6 + 4(n - 1)$ vertices, which is equal to $4n + 2$. Consequently, Theorem 6.2 holds. ∎

Theorem 6.3: The minimum number of vertices/corners in a grid of n hexagons in this study is given by $2n + 4$.

Proof: This Theorem can be proven by induction as well, where the first hexagon in the center of the grid contributes 6 vertices and each new hexagon next to it contributes at least 2 new vertices. And thus a grid of n equal hexagons will have at least $6 + 2(n - 1)$ vertices, which is equal to $2n + 4$. Consequently, Theorem 6.3 holds.   ∎

As we add more hexagons, and n goes to ∞, we find that the vast majority of the new hexagons will contribute 2 new vertices each. Specifically, the new hexagons that contribute more than 2 become exceedingly rare and there are never new ones contributing only 1 new vertex.

Next, we describe our deployment algorithm for RNs and CNs in the target area. In Algorithm 6.1, lines 1–5 describe the inputs required to come up with the deployment plan. The size of the target area, number of SNs available and their communication range, and the location of the sink are essential to decide on the outputs described in Steps 7 and 8. They are the numbers and positions of RNs and CNs required for maintaining connectivity of the SNs with the sink. In Step 9, the receiver sensitivity of the RNs and CNs is set to −101 dBm, which is typically the value in commercially available SNs and RNs. In Step 10, a threshold value of signal strength γ_{th} is set such that it is 3 dBm above R_{sense}. This is to guarantee reception of signals that are stronger than the receiver's sensitivity, which is the least value of the signal that it can detect. Once these values are set, we use Equation 6.18 to plot a graph of the variation of P_r as a function of d at transmit powers in the range (−5 to 10 dBm). The transmitter and receiver represent the RNs and CNs. In line 13, the values obtained from the plot are tabulated to strategically identify a value of d in Step 14 to ensure that there is at least one RN or CN lying between any two SNs to guarantee connectivity of SNs with the sink across the entire network. In line 15, we ensure that the transmit power for the chosen d is able to support $P_r > 0.8$ for every link, at least under near optimal conditions. Once the side of each hexagon grid is identified in Step 16, lines 17–20 describe the steps to identify the number of rows and columns in the hexagon grid covering the target area and numbers and positions of RNs and CNs in each grid cell. Thus, for SNs placed uniformly and randomly in a target region, Algorithm 6.1 gives the deployment plan for placing RNs and CNs in the area.

Algorithm 6.1: O2D Deployment Strategy for RNs and CNs

1. **Inputs:**

2.   Target hexagon area A

3.   Number of sensor nodes

4.   Sensor node communication range (r_{SN})

5.   Sink position at the center of target area

6. **Outputs:**

7.   Counts of RNs and CNs for the target area

8.   Positions of RNs and CNs in the deployment region

9. **Initialize:**

10.   Receiver sensitivity is set to R_{sense}

11.   Threshold signal strength γ_{th}

12. **Begin:**

13.   Plot P_r graph against d, for different transmit powers

14.   Tabulate values from the plot in Step 12

15.   Identify a value of d such that $d \leq 2*(r_{SN})$ and $P_r > 0.8$

16.   Choose P_t such that $0 \leq P_t \leq 10$, for $P_r > 0.8$ at d

17.   Set d as the side length L of each hexagon in A

18.   Approximate number of hexagon grids required to cover the target area using $A/(1.5*\sqrt{3}*L^2)$

19.   Round up this area to the nearest higher number G

20. **End**

Taking advantage of the foregoing knowledge of the SNs and grid vertices positions, RNs can be positioned on any grid vertex as long as they are within the probabilistic communication range of the largest amount of SNs and CNs. This can significantly limit the feasible search space without affecting the quality of the positioning strategy. To further elaborate on our strategy optimality, we introduce the following definitions:

> **Definition 6.1:** (Optimal Set): A finite set of positions P is optimal if and only if it satisfies the following property: There exists an optimal placement of RNs in which each available RN is placed at a position in P.

We aim at finding such an *optimal* set in order to achieve a more efficient discrete search space in which candidate relays' positions are not including all of the grid vertices but rather a subset of these vertices that has the most potential to enhance the network reliability. Moreover, since the computational complexity will be proportional to the cardinality of this optimal set P, we should find a set with a reasonably small size.

> **Definition 6.2:** (Covered Grid Unit (CGU)): A covered grid unit α is a grid unit that has a connected center with at least one CN. Let $C(\alpha)$ denote the subset of the CN coordinates covering α.

We assume that the considered virtual hexagon grid can have a building unit, called *grid unit*. For example, the *Grid Unit (GU)* of the 2-D grid shown in Figure 6.3 is the small hexagon drawn by dashed lines. Each *grid unit* has a center of mass represented by its position coordinates (black dots in Figure 6.3). We decide whether a *grid unit* is covered by a specific CN if the probabilistic connectivity between the *grid unit* center and that node is greater than or equal to the aforementioned threshold (P_r).

> **Definition 6.3:** (Maximal Covered Grid Unit (MCGU)): A covered grid unit α is maximal if there is no Covered Grid Unit β, where $C(\alpha) \subseteq C(\beta)$.

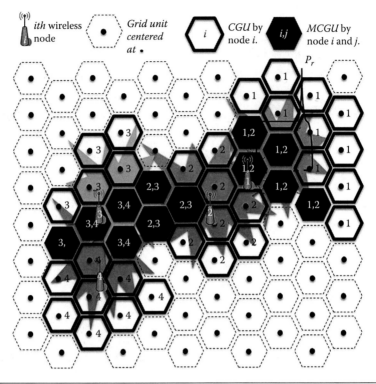

Figure 6.3 An example of a maximum covered grid unit in a 2-D plane. Numbers inside the bounded hexagons represent the node ID covering these hexagons. Arrows from the grid units' centers to the RN indicate the ability to communicate.

For more illustration, consider the 2-D plane shown in Figure 6.3 again. It shows four wireless nodes with respect to the grid units. Each wireless node has an arbitrary communication range. The covered grid units are bounded by solid lines and the MCGUs are solid black hexagons. These MCGUs have the highest potential to place the RNs due to their ability to establish the highest number of new edges between already deployed SNs. Accordingly, we have to show that an optimal set can be derived from the set of MCGUs. Towards this end, we state the following Lemmas.

Lemma 6.4: For every CGU β, there exists an MCGU α such that $C(\beta) \subseteq C(\alpha)$.

Proof: If β is an MCGU, we choose α to be β itself. If β is not an MCGU, then by definition, there exists a covered grid unit α_1 such that $C(\beta) \subseteq C(\alpha_1)$. If α_1 is an MCGU, we choose β to be α_1, and if α_1 is not maximal, then by definition, there exists another covered grid unit α_2 such that $C(\alpha_1) \subseteq C(\alpha_2)$. This process continues until a maximal covered grid unit α_x is found; we choose α to be α_x. Thus, Lemma 6.4 holds.   ∎

Lemma 6.5: Finding an MCGU takes at most $(n-1)$ step, where n is number of RNs.

Proof: By referring to the proof of Lemma 6.4, it is obvious that $|C(\alpha x)| \leq n$, and $|C(\alpha)| \langle |C(\alpha 1)| \langle |C(\alpha 2)| < \cdots < |C(\alpha x)| \leq n$; where $|C|$ represents the cardinality of the set C. Consequently, the process of finding the maximal covered grid unit α_x takes a finite number of steps less than or equal to $n-1$.   ∎

Accordingly, we can reach the following Theorem:

Theorem 6.4: A set P that contains one position from every MCGU is optimal.

Proof: To prove this Theorem, it is sufficient to show that for any arbitrary placement Z, we can construct an equivalent one $\bar{Z}$ in which every RN is placed at a position in P. In other words, the placement of an RN at position i, within the communication range of the SNs x and y, is equivalent to the placement of the same RN node at position j within the communication range of the SNs x, y, and z. Towards this end, let us assume that in Z, a relay node i is placed such that it is connected to a subset J of SNs. It is obvious that there exists a covered grid unit β, such that $J \subseteq C(\beta)$. From Lemma 6.4, there exists an MCGU α such that $C(\beta) \subseteq C(\alpha)$. In $\bar{Z}$, we place i at the position in P that belongs to α, so that i is placed at a position in P and is still connected with all SNs in J. By repeating this for all RNs, we construct a placement $\bar{Z}$ which is equivalent to Z, and thus, Theorem 6.1 holds.   ∎

6.4 Case Study

This case study elaborates more on the proposed O2D deployment strategy, and finds the following: (1) the grid edge length (L) and the distance (d) at which $P_r > 0.8$, and (2) the total number of required RNs and CNs to cover a hexagon area of edge length equal to 1050 m. In this use case, we consider the deployment plan for a WSN with 1500 sensor nodes (SNs), $R_{sense} = -101$ dBm, $r_{SN} = 175$ m, $\gamma_{th} = -98$ dBm, and a sink at the center. The SNs are distributed randomly and uniformly over the targeted hexagon. Cognitive nodes and relay nodes are deployed at fixed equidistant locations on a 2-D grid of hexagon vertices. The relay nodes are to be placed at the grid vertices and the cognitive nodes at the center of each hexagon. The node connections and interactions are based on a hierarchical ZigBee-based model [50], where we assume an IEEE 802.15.4 MAC-PHY cross layer model. Using our proposed O2D approach, we identify the grid size and recommend the communication distance for successful data reception. We plot the probability P_r in Equation 6.18 as a function of the separation distance "d" between the transmitter and receiver for transmission powers of –3, 0 and 5 dBm, while γ_{th} is equal to –101 dBm as shown in Figure 6.4. The threshold value is chosen based on receiver

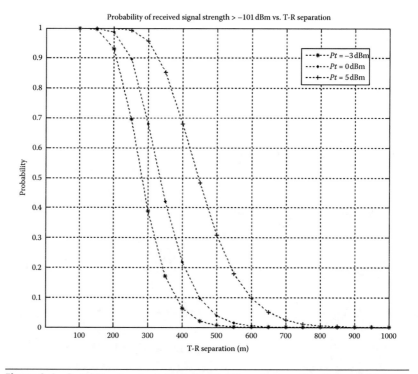

Figure 6.4 Probability of received signal strength (P_r) vs. the separation distance (d) between sender and receiver.

sensitivity values of the commercially available ZigBee transceivers, Atmel in this case [51]. From the graph in Figure 6.4, we can see that for a transmit power of $P_t = 0$ dBm, the probability of having received signal strength that is greater than -101 dBm is approximately 0.9 for a separation distance of about 250 m. Beyond this separation distance, we see that at 300 m, the probability drops to <0.7 for the same transmit power. However, if the transmit power is increased to $+5$ dBm, the communication range is almost 350 m for $P_r > 0.85$. Consequently, in Table 6.2, we tabulate the values of P_t and d for $P_r > 0.9$ and $P_r > 0.8$.

Accordingly, we find that at $P_t = 0$ dBm, a separation distance $d = 250$ m will be able to assure a receive power above -101 dBm at the receiver side when $P_r > 0.8$. Let us use this value of "d" to be the length of the hexagon side, and thus $d = L = 250$. The area covered by each hexagon grid will be $1.5 \times \sqrt{3} \times (250)^2$ since the target area of the bigger hexagon to be covered is equal to $1.5*\sqrt{3}*(1050)^2$. The total number of the small hexagon grids required to cover the target area is given by $1.5*\sqrt{3}*(1050)^2 / 1.5*\sqrt{3}*(250)^2$.

Then we round up this total number of small hexagons to the nearest higher whole number G, such that it becomes the recommended count of CNs in the network. Every CN is placed in the center of a hexagon cell. According to Theorems 6.2 and 6.3, there will be at most $(4G + 2)$ grid vertices utilized by RNs and at least $(2G + 4)$ vertices used by RNs as well. This will be the number of relay nodes at the corners of the assumed grid in the targeted area. Consequently, the goals stated at the beginning of this case study have been achieved.

In comparison to other kind of grids, such as the square grid for instance, we find that hexagon grids are more competitive because

Table 6.2   Values of "d" for Different Transmit Powers through Which We Can Achieve $P_r > 0.8$ and $P_r > 0.9$

$P_r > 0.9$		$P_r > 0.8$
P_t (dBm)	d (m)	d (m)
-3	150	190
0	200	225
3	250	280
5	275	300
7	300	350
10	360	400

the total number of vertices where we position the RNs in this study can be derived from the total number of small squares' rows in the big square grid as shown in Figure 6.2. Since the total number of small square cells is equal to G per our case study in this section, then the total number of rows is equal to $\sqrt{G}$. And thus the total number of vertices is equal to $(\sqrt{G} + 1)^2$. This by definition will be the number of RNs at the vertices of the square grid in the targeted area. Accordingly, the provided count of RNs in the hexagon grid scenario is at least twice the maximum possible count of RNs on the square grid. And thus, the count of RNs with respect to CNs will be much more than the achieved ratio in case of a square grid deployment. This can lead to an overall reduction in the cost the same as an MCGU can be achieved with more RNs (or cheap nodes) and less CNs (or expensive nodes) using the hexagon grid-based deployment.

6.5 Simulation Results and Discussion

The simulation results presented in this section are from Omnet++ and MATLAB® simulations. Our initial setup in Omnet++ considered the use of a planned 2-D grid deployment of CNs in a randomly deployed sensor network. No relay nodes were considered in this deployment. The grids were squares of side 200 m each. The CN transmit power varies between 0 and 10 dBm, with a specific increment value of {0, 3, 5, 7, 10} dBm. This means that CNs can communicate across the diagonals of the grid by increasing their transmit power, as described in Table 6.1, to achieve the desired communication range. The deployment plan is represented by Figure 6.5. We considered two techniques for data routing. The first technique (Technique A) is the directed diffusion (DD) technique [53], where the data was returned along the same path along which it arrived. The original technique was applied to sensor nodes. We apply the same strategy to CNs, with a restriction that the packets can travel only along the sides of the square grid to maintain the same transmit power at all CNs; this would be the case with the sensor nodes if they adopted the DD routing technique.

The second technique (Technique B) allowed the CNs to choose a different data delivery path, compared to the request arrival path. We also let the CNs communicate across the diagonals of the

Network topology with sensor nodes, local congnitive nodes and global cognitive node

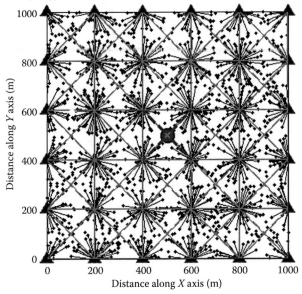

Figure 6.5   ICSN with only CNs and SNs in the network. Middle circle indicates sink or global cognitive node (GCN), small black diamonds indicate SNs, black triangles indicate CNs.

square grid. We let the simulations run till the first CN death for both the techniques.

From the simulations, we found that Technique B is better than Technique A in terms of network lifetime (time to first node death) and the remaining energy at CNs after first node death. This indicated that the network would be able to provide for a more graceful degradation of the network. Using Technique B, we see that network lifetime depends on the application and/or user requirements. In other words, network lifetime depends on the frequency with which requests arrive in the network. ICSNs are expected to support multiple applications [53], thus the lifetime of the network is completely dependent on the type and frequency of requests being served. From these simulations, we also found that communication across the diagonals of the square grid was more effective compared to the Manhattan routing in terms of power consumption and number of hops to reach the sink from the edge of the network. However, the transmit power had to be increased to ensure reliability of data arriving at the receiving node. And this would add to the total hardware cost of the network. Hence, we decided to make use of RNs to forward the data, as the nodes at the center of the grid were essentially being deployed for the purpose of increasing the communication range while decreasing the transmit power. While the cost is now higher than using just CNs in the network, it provides better connectivity at lower transmit powers of the CNs. Then, we further experimented with swapping the position of the RNs with the CNs in order to maintain the same level of connectivity. We also reduced the cost of the network hardware by increasing the number of RNs and decreasing the number of CNs used in the network. This deployment is as shown in Figure 6.6. As elaborated in Equation 6.16, the cost of a CN is greater than the cost of an RN, and thus this strategy would help us achieve better cost and energy performance.

In the subsequent sections, we describe the simulation setup for our experiments with the deployment plan implemented using MATLAB to allow more flexibility with the parameter setting and to control the CN behavior during network operation. The goal for these simulations was to minimize the number of CNs used in the network and to evaluate the deployment plan in terms of the QoI requirements.

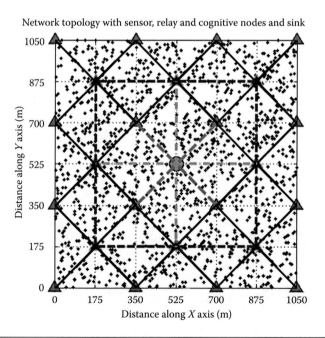

Network topology with sensor, relay and cognitive nodes and sink

Figure 6.6 ICSN with CNs, RNs, and SNs in the network. Middle circle indicates sink, small black diamonds indicate SNs, black triangles indicate CNs, gray triangles indicate RNs.

6.5.1 Simulation Setup

Using MATLAB (R2013a), we first simulated the deployment plan for a large-scale WSN with 1500 SNs, 16 RNs, 9 CNs, and a sink over a square and hexagon target area of a side length $A = 1050$ m. Later, we varied the CNs/RNs counts for more comprehensive analysis. The SNs were distributed randomly and uniformly over the target area. CNs and RNs are deployed at fixed equidistant locations on a 2-D square and hexagon grids as described in the aforementioned deployment plan. The RNs are deployed at the corners of each square/hexagon grid and the CNs are at the center. The sink is deployed at the center of the deployment region. The node connections and interactions are based on a hierarchical ZigBee topology model [15]. The network is built over an IEEE 802.15.4 MAC-PHY simulator based on the work in Reference [50–52]. We make use of the parameters listed in Table 6.3 for the MATLAB simulations.

In the following subsection, we evaluate the square grid deployment strategy in terms of QoI-based attributes such as the node reliability (NR) and the instantaneous throughput (IT).

6.5.2 Evaluation of the Square-Based Grid

We use delay, reliability, and instantaneous throughput as QoI evaluation metrics for the proposed deployment strategy.

Table 6.3   Simulation Parameters

PARAMETER	VALUE
Operational frequency	916 MHz (ISM band)
Data rate	250 kbps
Transmit power	0 dBm for RNs Varying from 0 to 10 dBm for CNs
Modulation	Amplitude shift keying
Encoding	Non-return to zero
Path loss model	Log normal shadowing; $n = 4$, $\sigma = 4$
SN transmit range	175 m
Payload size	0–127 bytes
Per node offered load	0.1–1.4 times the application payload

6.5.2.1 Node Reliability (NR)   Node reliability is defined as the probability that a transmitting node is able to successfully deliver a data packet to its next hop neighbor. It is a function of the node's buffer capacity, represented by the blocking probability, and the channel conditions at the time of channel access/data transmission. Thus, it inherently reflects the communication link reliability as well. This definition of reliability is based on the proposed work in Reference [50] for low power nodes in the 802.15.4 MAC-PHY model as follows:

$$NR = \left(\left(1 - P_{blocking} \right) * \left(1 - P_{c.\,fail} \right) * \left(1 - P_{p.discard} \right) \right) \qquad (6.21)$$

where $P_{blocking}$ represents the blocking probability due to a buffer full condition; $P_{c.\,fail}$ is the channel access failure probability and $P_{p.discard}$ is the probability that a packet is discarded on reaching the maximum number of retransmissions.

6.5.2.2 Instantaneous Throughput (IT)   The definition for instantaneous throughput (IT) at a receiving node is based on the work in Reference [51], and is applied to both relay and cognitive nodes. It is defined as the ratio of the size of the frame payload at the physical layer L in bits, to the mean service time M in seconds.

$$IT = L/M \, (\text{bits/s}) \qquad (6.22)$$

where the payload size L consists of both, the overhead and the application payload size.

6.5.2.3 Delay   The observed average waiting time at a receiving node accounts for delays due to the mean service time at the transmitting node, which is a function of the frame arrival rate. Along with the variation in these three QoI attributes, Figures 6.7 through 6.9 examine the proposed deployment strategy as the frame arrival rates increase. Figure 6.7 indicates that the instantaneous throughput drops to almost half its value of about 5×10^4 bps for a frame arrival rate of 6 fps, from an original value of 9.8×10^4 bps for a frame arrival rate of less than 1 fps. The average wait time also sees a steeper increase beyond a 5 fps frame arrival rate. From Figure 6.8, we can see that as the frame arrival rate increases beyond 6 fps, the reliability drops below 0.9 and the average delay for successfully received

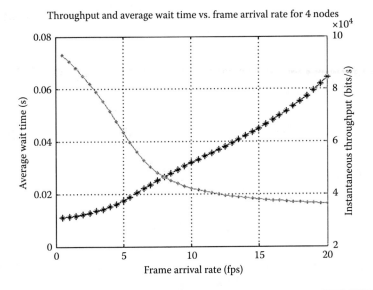

Figure 6.7   Average wait time and throughput vs. per node frame arrival rate.

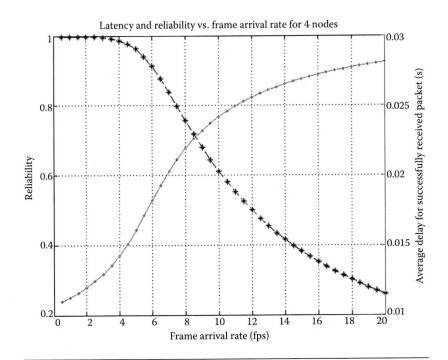

Figure 6.8   Reliability and average delay for successfully received packets vs. per node frame arrival rate.

packets is about 75% more from about 0.01 s at 1 fps arrival rate to 0.018 s at 6 fps. From the results in Figure 6.9, we draw the inference that a drop in reliability could be an indicator of poor node or link performance or both. It can be seen that IT is more affected by variations in the offered load when compared with NR for the same deployment strategy. In WSNs deployed for USN applications, the cognitive nodes can be utilized to identify the end user's service requirements and manage the node associations and resources accordingly. If an application requires high throughput guarantees, then cognitive nodes can exercise their learning mechanism to understand the user requirements. And the reasoning mechanism can exercise its control to limit the number of nodes scheduled for simultaneous transmission, to provide the user desired reliability for the application being serviced. These control instructions from cognitive nodes are passed on to RNs as well; the performance of the entire network can be tuned to meet the application's service requirements.

In the following subsection, we evaluate the hexagon grid deployment strategy in terms of QoI-based attributes such as NR and IT.

6.5.3 Evaluation of the Hexagon-Based Grid

We make use of the above mentioned simulation setups and parameters in Table 6.3, while varying the CN count between 10 and 60 and

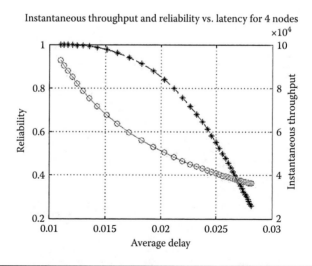

Figure 6.9 Reliability and throughput vs. average delay for successfully received packets.

the RNs to be more than 4 and less than 6 per CN in assessing the *Hexagon grid* deployment.

6.5.3.1 Instantaneous Throughput (IT)   For IT assessment, the load offered by a transmitting node was varied from 1 to 1500 bps and the *IT* seen at a receiving node was recorded for different counts (N) of cognitive nodes ranging from 10 to 60 nodes. According to results shown in Figure 6.10, an overall trend of decline in IT is observed while increasing the offered load and N values. For a given value of N, the throughput decreases as the offered load increases. For instance, at N equal to 20, IT drops by about 33% from a value of 120 kbps at ~100 bps to <80 kbps when the offered load is increased from 1 to 1500 bps. However, at a higher value of N equal to 60, the percentage drop in IT is about 65% for the same increment in the offered load. Furthermore, we observe better IT performance in comparison to the IT achieved while applying the square grid-based deployment. For example, while experiencing the worst IT in Figure 6.10 when N is equal to 60, it is still better than the IT performance depicted in Figure 6.7 when every frame consists of 100 bits.

Furthermore, results in Figure 6.11 show the variation of the IT as a function of the offered load, while examining different transmission powers. Obviously, as the transmit power increases, the IT performance improves. In addition, it degrades while the traffic load per node increases.

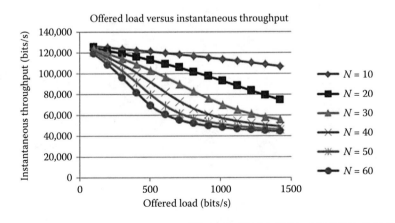

Figure 6.10   Offered load vs. the instantaneous throughput.

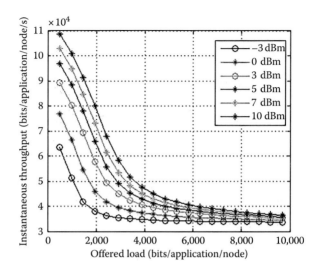

Figure 6.11   Offered load vs. the IT at different transmit powers.

6.5.3.2 Node Reliability (NR)   We apply the aforementioned NR definition while varying the CN count, as they will be interacting with sensor and relay nodes placed on the grid at positions identified by the deployment strategy. The load offered per RN/CN was varied and its impact on the *NR* was observed. In this simulation part, we examine the cognitive node's performance in terms of reliability by varying their counts (*N*) as shown in Figure 6.12. From Figure 6.12, it can be seen that *NR* does not drop below 0.9 even at the maximum offered

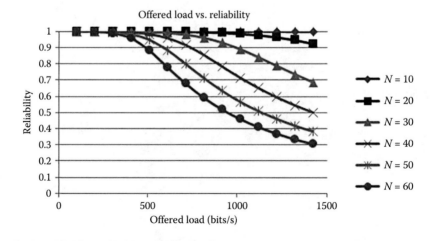

Figure 6.12   Offered load vs.reliability for data rate of 250 kbps.

load of 1500 bps when the value of N is 20 or less. At N equal to 30, a linear drop in NR is seen when the offered load increases beyond 1000 bps. However, when the value of N is increased beyond 30, we notice that for an offered load of 750 bps, NR drops from approximately 1 at N equal to 30 to below 0.7 at N equal to 60 cognitive nodes. Again in comparison to Figure 6.8 while applying the square grid solution, the hexagon-based deployment performs much better in terms of reliability.

Moreover, the results in Figure 6.13 show the variation of the NR (i.e., reliability) while increasing the per node offered load, for different values of the transmit power.

From these results, it can be said that even if the nodes are strategically deployed for reliable reception of data, if the number of cognitive nodes is >30 when the per node offered load is >500 bps, NR drops below 0.9. Thus it is important to assure that unnecessary cognitive nodes are in sleep mode to maximize the benefits of the deployment strategy in terms of reliability.

According to simulation results in Figures 6.10 and 6.12, it can be seen that the IT is more affected by variations in offered load and values of N when compared with the NR for the same deployment strategy.

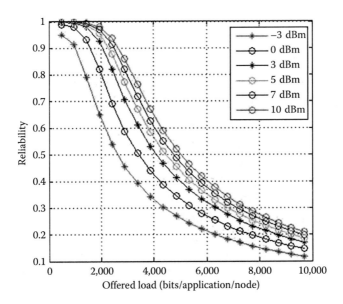

Figure 6.13   Offered load vs. NR at different transmit powers.

In WSNs deployed for ubiquitous sensing applications, the cognitive nodes can be utilized to identify the end user's service requirements and manage the node associations and resources accordingly. If an application requires high throughput guarantees, then CNs can exercise their learning mechanism to understand the user requirements and the reasoning mechanism can exercise its control to limit the number of awake CNs and to provide the user desired reliability for the application being serviced. These control instructions from CNs are passed on to RNs as well and the performance of the entire network can be tuned to meet the application's service requirements.

Moreover, the CN can make use of the observed NR values at each node to deliver data to the user with high reliability along each hop towards the sink. Thus, the combined use of a deployment strategy and the cognitive elements in the CN in ubiquitous sensing applications can be used to deliver data with user desired values on the data attributes of NR and IT.

6.6 Conclusions

WSNs require additional capabilities in terms of being able to understand the dynamic behavior and store the large volume of generated data in the USN infrastructure. The network is dynamic due to changing application requirements, end user behavior, and network topology. To address these challenges and meet the changing application requirements, we proposed the use of CNs which implement learning, reasoning, and knowledge representation in WSNs for USN applications. The reasoning mechanism helps to respond to current network conditions that require immediate attention. The learning mechanism, along with the knowledge acquired during the network operation, is used for planning and controlling the response to predicted network conditions. The deployment planning will help to use the CNs more effectively in exercising the learning and reasoning mechanisms. Thus, a deployment plan for CNs in large scale WSNs for USN applications was strategized; the performance of two attributes, node reliability and instantaneous throughput, were evaluated. We found that the strategy was successful in satisfying the end user's requirements when the capabilities of the CNs are effectively used to exercise control over the other nodes in the network. The cloud computing storage infrastructure helps to store and process the data generated within the network. In our future work, we will explore the routing and caching capabilities of the cognitive nodes to further expand their applications in various USN applications in the cloud computing era.

References

1. ITU-T technology watch briefing report series, no. 4, Ubiquitous sensor networks, February 2008.
2. F. Al-Turjman, Information-centric sensor networks for cognitive IoT: An overview, *Springer Annals of Telecommunications Journal*, vol. 72, no. 1, pp. 3–18, 2017.
3. I. Stojmenović and S. Olariu, Data-centric protocols for wireless sensor networks, *Handbook of Sensor Networks*, I. Stojmenović, ed., John Wiley & Sons, Hoboken, NJ, pp. 417–456, 2005.
4. M. Z. Hasan, F. Al-Turjman, and H. Al-Rizzo, Evaluation of a duty-cycled protocol for TDMA-based wireless sensor networks, In *Proceedings of the International Wireless Communications and Mobile Computing Conference*, Paphos, Cyprus, 2016, pp. 964–969.
5. S. Oteafy, A framework for heterogeneous sensing in big sensed data, *IEEE Global Communications Conference (GLOBECOM)*, Washington, DC, 2016, pp. 1–6.
6. S. Oteafy and H. Hassanein, Resilient IoT architectures over dynamic sensor networks with adaptive components, *IEEE Internet of Things Journal*, vol. 4, no. 2, pp. 474–483, 2017.
7. A. Al-Fagih, F. Al-Turjman, W. Alsalih, and H. Hassanein, A priced public sensing framework for heterogeneous IoT architectures, *IEEE Transactions on Emerging Topics in Computing*, vol. 1, no. 1., pp. 133–147, 2013.
8. F. Al-Turjman, A. Alfagih, W. Alsalih, and H. Hassanein, A delay-tolerant framework for integrated RSNs in IoT, *Elsevier: Computer Communications Journal*, vol. 36, no. 9, pp. 998–1010, 2013.
9. A. Alfagih, F. Al-Turjman, and H. Hassanein, Ubiquitous robust data delivery for integrated RSNs in IoT, In *Proceedings of the IEEE Global Communications Conference (GLOBECOM'12)*, Anaheim, CA, 2012, pp. 298–303.
10. P. Zhang, Z. Yan, and H. Sun, A novel architecture based on cloud computing for wireless sensor network, In *Proceedings of the 2nd International Conference on Computer Science and Electronics Engineering (ICCSEE)*, Atlantis Press, Paris, France, 2013.
11. W. Kurschl and W. Beer, Combining cloud computing and wireless sensor networks, In *Proceedings of the 11th International Conference on Information Integration and Web-Based Applications and Services (iiWAS'09)*, ACM, New York, 2009, pp. 512–518.

12. Z. Yun, X. Bai, D. Xuan, T. H. Lai, and W. Jia, Optimal deployment patterns for full coverage and k-connectivity ($k \le 6$) wireless sensor networks, *IEEE/ACM Transactions on Networking (TON)*, vol. 18, no. 3, pp. 934–947, 2010.

13. P. Cheng, C. N. Chuah, and X. Liu, Energy-aware node placement in wireless sensor networks, In *Global Telecommunications Conference, 2004. GLOBECOM'04*, IEEE, Dallas, TX, 2004, vol. 5, pp. 3210–3214.

14. M. Cardei, M. T. Thai, Y. Li, and W. Wu, Energy-efficient target coverage in wireless sensor networks, In *Proceedings IEEE 24th Annual Joint Conference of the IEEE Computer and Communications Societies, INFOCOM 2005*, Miami, FL, 2005, vol. 3, pp. 1976–1984.

15. F. M. Al-Turjman, A. E. Al-Fagih, H. S. Hassanein, and M. Ibnkahla, Deploying fault-tolerant grid-based wireless sensor networks for environmental applications, In *2010 IEEE 35th Conference on Local Computer Networks (LCN)*, IEEE, Denver, CO, 2010, pp. 715–722.

16. L. Tran-Thanh and J. Levendovszky, A novel reliability based routing protocol for power aware communications in wireless sensor networks, In *Proceedings of the 2009 IEEE Wireless Communications and Networking Conference (WCNC)*, April 05–08, 2009, pp. 2308–2313.

17. A. Tufail, Reliable latency-aware routing for clustered WSNs, *International Journal of Distributed Sensor Networks*, vol. 2012, 6 p., Article ID. 681273, 2012.

18. ZigBee specifications. www.zigbee.org ZigBee Document 053474r17, 2008.

19. X. Cheng, D. Z. Du, L. Wang, and B. Xu, Relay sensor placement in wireless sensor networks, *ACM/Springer Journal of Wireless Networks*, vol. 14, no. 3, pp. 347–355, 2008.

20. X. Han, X. Cao, E. L. Lloyd, and C. C. Shen, Fault-tolerant relay node placement in heterogeneous wireless sensor networks, *IEEE Transactions on Mobile Computing*, vol. 9, no. 5, pp. 643–656, 2010.

21. E. L. Lloyd and G. Xue, Relay node placement in wireless sensor networks, *IEEE Transactions on Computers*, vol. 56, no. 1, pp. 134–138, 2007.

22. K. Xu, H. Hassanein, G. Takahara, and Q. Wang, Relay node deployment strategies in heterogeneous wireless sensor networks, *IEEE Transactions on Mobile Computing*, vol. 9, no. 2, pp. 145–159, 2010.

23. F. Al-Turjman, H. Hassanein, and M. Ibnkahla, Optimized relay placement to federate wireless sensor networks in environmental applications, In *2011 7th International Wireless Communications and Mobile Computing Conference (IWCMC)*, July 4–8, 2011, pp. 2040–2045.

24. M. Li and Y. Liu, Underground coal mine monitoring with wireless sensor networks, *ACM Transactions on Sensor Networks*, vol. 5, no. 2, 29 p., Article 10, 2009.

25. F. Al-Turjman, Hybrid approach for mobile couriers election in smart-cities, In *Proceedings of the IEEE Local Computer Networks (LCN)*, Dubai, UAE, 2016, pp. 1–4.

26. F. Al-Turjman, Cognitive routing protocol for disaster-inspired internet of things, *Elsevier Future Generation Computer Systems*, doi:10.1016/j.future.2017.03.014, 2017.

27. F. Wang, D. Wang, and J. Liu, Traffic-aware relay node deployment: Maximizing lifetime for data collection wireless sensor networks, *IEEE Transactions on Parallel and Distributed Systems*, vol. 22, no. 8, pp. 1415–1423, 2011.

28. M. Z. Hasan and F. Al-Turjman, Evaluation of a duty-cycled asynchronous X-MAC protocol for vehicular sensor networks, *EURASIP Journal on Wireless Communications and Networking*, doi:10.1186/s13638-017-0882-7, 2017.

29. Y. B. Reddy and C. Bullmaster, Application of game theory for cross-layer design in cognitive wireless networks, In *Proceedings of the 6th International Conference on Information Technology: New Generations, ITNG*, IEEE, Washington, DC, 2009, pp. 510–515.

30. L. Reznik and G. Von Pless, Neural networks for cognitive sensor networks, In *Proceedings of IEEE International Joint Conference on Neural Network (IJCNN)*, IEEE, Hong Kong, China, 2008, pp. 1235–1241.

31. C. Bisdikian, L. M. Kaplan, and M. B. Srivastava, On the quality of information in sensor networks, *ACM Transactions on Sensor Networks*, vol. 9, no. 4, Article 48, 2013.

32. G. Singh and F. M. Al-Turjman, Learning data delivery paths in QoI-aware information-centric sensor networks, *IEEE Internet of Things Journal*, vol. 3, no. 4, pp. 572–580, 2016.

33. B. Ahlgren, C. Dannewitz, C. Imbrenda, D. Kutscher, and B. Ohlman, A survey of information-centric networking, *IEEE Communications Magazine*, vol. 50, no. 7, pp. 26, 36, 2012.

34. F. Al-Turjman and H. Hassanein, Enhanced data delivery framework for dynamic information-centric networks (ICNs), In *Proceedings of the IEEE Local Computer Networks (LCN)*, Sydney, Australia, 2013, pp. 831–838.

35. B. Krishnamachari, D. Estrin, and S. Wicker, Modelling data-centric routing in wireless sensor networks, *IEEE INFOCOM*, vol. 2, pp. 39–44, 2002.

36. F. Al-Turjman, Cognitive caching for the future: Fog networking, *Elsevier Pervasive and Mobile Computing*, doi:10.1016/j.pmcj.2017.06.004, 2017.

37. W. R. Heinzelman, A. Chandrakasan, and H. Balakrishnan, Energy-efficient communication protocol for wireless microsensor networks, In *IEEE Proceedings of the 33rd Hawaii International Conference on System Sciences*, IEEE, Maui, HI, 2000, pp. 1–10.

38. F. Al-Turjman, Cognitive-node architecture and a deployment strategy for the future sensor networks, *Springer Mobile Networks and Applications*, doi:10.1007/s11036-017-0891-0, 2017.

39. Q. Zhao, L. Tong, and Y. Chen, Energy-aware data-centric MAC for application-specific sensor networks, In *2005 IEEE/SP 13th Workshop on Statistical Signal Processing*, July 17–20, 2005, pp. 1238–1243.

40. G. Vijay and M. Ibnkahla, CCAWSN: A cognitive communication architecture for wireless sensor networks, In *26th Biennial Symposium on Communications (QBSC)*, May 28–29, 2012, pp. 132–137.

41. G. Vijay, E. B. Bdira, and M. Ibnkahla, Cognition in wireless sensor networks: A perspective, *IEEE Sensors Journal*, vol. 11, no. 3, pp. 582, 592, 2011.

42. M. Z. Hasan, H. Al-Rizzo, and F. Al-Turjman, A survey on multipath routing protocols for QoS assurances in real-time multimedia wireless sensor networks, *IEEE Communications Surveys and Tutorials*, doi:10.1109/COMST.2017.2661201, 2017.

43. Lithium based batteries. http://batteryuniversity.com/learn/article/lithium_based_batteries.

44. M. Zuniga and B. Krishnamachari, Analyzing the transitional region in low power wireless links, In *Sensor and Ad Hoc Communications and Networks, First Annual IEEE Communications Society Conference on, IEEE SECON*, Santa Clara, CA, pp. 517–526, October 2004.

45. F. Al-Turjman, Price-based data delivery framework for dynamic and pervasive IoT, *Elsevier Pervasive and Mobile Computing Journal*, doi:10.1016/j.pmcj.2017.05.001, 2017.

46. L. Bölöni and D. Turgut, Value of information based scheduling of cloud computing resources, *Future Generation Computer Systems Journal (Elsevier)*, vol. 71, pp. 212–220, 2017.

47. D. Turgut and L. Bölöni, Value of information and cost of privacy in the internet of things, *IEEE Communications Magazine*, vol. 1, pp. 9–52, 2017.

48. T. S. Rappapport, *Wireless Communications: Principles and Practice*, Prentice Hall, Upper Saddle River, NJ, 2001.

49. F. Al-Turjman, H. Hassanein, and M. Ibnkahla, Quantifying connectivity in wireless sensor networks with grid-based deployments, *Journal of Network and Computer Applications*, vol. 36, no. 1, pp. 368–377, 2013.

50. M.-H. Zayani and V. Gauthier, Usage of IEEE 802.15.4 MAC–PHY Model. www-public.it-sudparis.eu/~gauthier/Tools/802_15_4_MAC_PHY_Usage.pdf.

51. M.-H. Zayani, V. Gauthier, and D. Zeghlache, A joint model for IEEE 802.15.4 physical and medium access control layers, In *Proceedings of IEEE the 7th International Wireless Communications and Mobile Computing Conference (IWCMC)*, IEEE, Istanbul, Turkey, 2011.

52. C. Intanagonwiwat, R. Govindan, and D. Estrin, Directed diffusion: A scalable and robust communication paradigm for sensor networks, In *Proceedings of the 6th Annual International Conference on Mobile Computing and Networking*, ACM, Boston, MA, 2000.

53. M. Z. Hasan, F. Al-Turjman, and H. Al-Rizzo, Optimized multi-constrained quality-of-service multipath routing approach for multimedia sensor networks, *IEEE Sensors Journal*, vol. 17, no. 7, pp. 2298–2309, 2017.

54. P. Park, P. Di Marco, P. Soldati, C. Fischione, and K. H. Johansson, A generalized Markov chain model for effective analysis of slotted IEEE 802.15.4, In *IEEE 6th International Conference on Mobile Adhoc and Sensor Systems, MASS'09*, IEEE, Macau, China, 2009, vol. 130, no.139, pp. 12–15.

7

TOWARDS PROLONGED LIFETIME FOR DEPLOYED WIRELESS SENSOR NETWORKS IN OUTDOOR ENVIRONMENT MONITORING*

Wireless networking and advanced sensing technology have enabled the development of low cost, power efficient Wireless Sensor Networks (WSNs) that can be used in various application domains such as healthcare, military, and OEM [1]. The main building blocks of WSNs are Sensor Nodes (SNs). These nodes collect information, i.e., sense and some physical and/or chemical properties of a monitored environment and transmit these measurements to a central node known as the Base Station (BS). This transmission can either be periodic or on demand [2]. Among the various application domains of WSNs, Outdoor Environment Monitoring (OEM) attracted considerable attention due to its unique characteristics [3,4]. These include large monitored areas, isolated and distant territories, harsh operational conditions, and high probabilities of node and link failures [5,6]. Fortunately, a well-planned WSN deployment in such environments offers a reliable, yet cheap, means of decentralized data collection with minimal human intervention. However, in order to maintain a prolonged and reliable monitoring, the WSN needs to withstand the harsh operational conditions of outdoor environments like heavy rains, snowfalls, sandstorms, extreme temperature variations, etc. These conditions may cause a significant percentage of node and link failure [7,8], which can be captured and statistically quantified using the Probability of Node Failure (PNF) and Probability of

* This chapter has been coauthored with Hossam S. Hassanein and Mohamad Ibnkahla.

143

Disconnected Nodes (PDN). Hence, a well-planned WSN deployment should reduce these two probabilities [9]. While PNF can be partially reduced through proper placement and packaging of the nodes, PDN can significantly be reduced by a fault-tolerant deployment [10]. Fault-tolerance is a pivotal step for sustained and reliable monitoring. It is achieved through injecting a certain level of redundancy in the network such that it can withstand a given percentage of failure while maintaining the desired monitoring level.

Due to the scarcity of energy resources in outdoor environments, network nodes are almost always battery powered. However, since a WSN in an OEM application is envisioned to work unattended for long periods of time, a stringent constraint is imposed on the energy consumption per node. This becomes a serious challenge when the network monitoring area is huge. In this case, energy can be drastically consumed if a distant BS is to be reached directly by all the SNs with a blind knowledge of the remaining energy budget per node in the network. To overcome this, a distributed system of Relay Nodes (RNs)/SNs can be used [11]. An RN is a dedicated communication node with larger energy storage, i.e., a larger battery than regular SNs, capable of collecting data from a cluster of SNs and passing it to the BS. RNs can either be static or mobile. Unlike Static RNs (SRNs) that are located once within the network lifetime, Mobile RNs (MRNs) are given certain mobility features such that they can be relocated on demand [12]. The use of this type of RN helps resolve bottleneck problems during the network lifetime. In fact, MRNs can be seen as a proactive solution to maintain connectivity and fault-tolerance when some communication paths are running out of energy or losing connectivity [13]. Also, one of the unique features of OEM applications is the 3-D space monitoring where the height of a node is as important as its horizontal position [14], which cannot be considered by 2-D deployment algorithms. For instance, in monitoring the gigantic redwood trees in California, some experiments required placing the sensors at varying heights ranging from the ground surface up to tens of meters [15]. Moreover, monitoring the intensity of certain gases, like CO_2 [11,16,17], requires sensor placement at different heights such that monitoring coverage and accuracy requirements are met. Such a 3-D monitoring can easily be secured if the coverage space is modeled as a 3-D grid, which is a typical coverage model in

OEM applications [1,17]. Hence, network nodes can only be placed at the vertices of this grid. In fact, the grid model limits the search space to a finite number of points. The shape of the grid building units can either be a cube, a hexagon, an octahedron, or any regular shape chosen to meet certain coverage levels [18]. Other advantages of the 3-D grid modeling include exclusion of all positions where node deployment is not possible and accurate description of the possible routing paths [19]. In spite of the aforementioned grid advantages, placing the WSN nodes on the grid vertices might affect their deployment optimality. Nevertheless, this effect is fortunately controllable by the grid edge length (i.e., the deployment optimality is proportional with the count of vertices and inversely proportional with the grid edge length). Thus, a more restricted search space (without affecting the deployment optimality) is required. Meanwhile, the overall network cost is proportional to the total number of nodes deployed. Hence, the lower the number of nodes, the lower the overall cost. However, communication reliability and fault-tolerance require abundant node deployment. A tradeoff exists between the overall cost and the network performance. Therefore, the network deployment problem can be best modeled as an optimization problem. The objective is to maximize the network lifetime by reducing the energy consumption; the constraints are cost efficiency, communication reliability, and fault-tolerance. This problem will be mathematically modeled in subsequent sections.

7.1 Related Work

Recently, there have been several proposals for an energy efficient, lifetime maximizing WSN deployment. These proposals differ both in the type of nodes used in the network and in their deployment strategies. In terms of the type of nodes used, most of the work in literature has considered the use of only SNs deployed in a target area. Very few researchers have considered the use of RNs to improve the communication range and network lifetime. PEDAP and its power-aware version, PEDAP-PA [20], L-PEDAP [21], EESR [22], and MLDA [23] are examples of WSNs that consider only SNs deployed in a target area and implement various schemes to improve the network lifetime. For instance, PEDAP considers minimizing the total energy expended by the network in a round of communication, but it doesn't consider the issue of balancing energy consumption among the nodes. It consumes less energy to find a route and is able to achieve a good lifetime for the last node, but it doesn't provide for load balancing among SNs nor reliable communication in the network. PEDAP-PA is an improvised version of PEDAP that considers balancing the energy consumption among the SNs by computing their remaining energy using a cost function. However, this cost function considers only the transmitting nodes' residual energy. The routing tree is recomputed after a predefined number of rounds, which is a major drawback when considering an improvement in system reliability. Moreover, both PEDAP and PEDAP-PA are centralized algorithms that were designed for smaller deployment areas and might be unsuitable for large-scale deployments such as OEM applications. EESR and L-PEDAP consider the remaining energy levels at both the transmitter and transceiver SNs to achieve better load balancing. EESR uses Kruskal's algorithm for the routing tree construction; it works best when the SN is in the same transmission range and can communicate directly with the sink. L-PEDAP is capable of automatically rerouting a packet to the destination when it finds that the energy level of a node is less than the threshold value. Thus L-PEDAP achieves both load balancing and reliable communication as it is capable of identifying node failure and recovering from it, unlike EESR. However, L-PEDAP fails to minimize the energy consumption and communication time. This is true even in the case of MLDA that fails to achieve the desired tradeoff between communication delay and network lifetime in the system. Thus, none of these works in existing literature have been able to satisfactorily address the problem of maximizing network lifetime while reducing the energy consumption and while also considering the constraints of cost efficiency, communication reliability, and fault-tolerance.

Moving on to node deployment strategies, they are mainly classified as random and deterministic strategies (see Table 7.1). In random deployments, nodes' positions can be chosen in a purely random deployment plan or can be based on a weighted random deployment plan, where the distributed nodes' density is not uniform in the monitored areas. For instance, K. Xu et al. [24], studied random deployment of static RNs in a 2-D plane. The authors proposed an efficient

Table 7.1 A Comparison between Various Deployment Proposals in the Literature

| REFERENCE | CONSIDERED PERFORMANCE METRICS[a] | | | | DEPLOYMENT APPROACH | | |
	COST	CONNECTIVITY	FAULT-TOLERANCE	LIFETIME	TYPE	CENTRALIZED/ DECENTRALIZED	TARGETED SPACE
[9]	✓	✓	—	—	Deterministic	Decentralized	2-D
[25]	✓	✓	—	✓	Deterministic	Decentralized	2-D
[16]	—	✓	✓	—	Random	Centralized	2-D
[24]	✓	✓	—	—	Random	Centralized	2-D
[26]	✓	✓	—	✓	Deterministic	Centralized	2-D
[27]	—	✓	—	✓	Deterministic	Centralized	2-D
[28]	—	✓	✓	—	Random	Centralized	3-D
[29]	✓	✓	✓	—	Deterministic (grid-based)	Centralized	2-D
[30]	—	✓	✓	✓	Deterministic	Centralized	2-D
Our work	✓	✓	✓	✓	Deterministic (grid-based)	Decentralized	3-D

[a] The √ in this column indicates that the corresponding metric is considered. The — means the metric is not considered.

deployment strategy that maximizes the network lifetime when all RNs reach the BS with a single hop only. Motivated by the weakness of the uniform random deployment, the authors proposed a weighted random deployment strategy with a gradually increasing density of nodes as the distance to the BS increases. This strategy compensates the number of RNs for the energy needed to reach the BS. Hence, monitoring reliability can be sustained for longer periods of time, i.e., network lifetime is maximized.

In contrast, deterministic deployments aim at deploying nodes exactly on specific, predefined locations. These deployments can be accomplished through *centralized* or *distributed* approaches (see Table 7.1). In centralized approaches, global information gathering is required to end up with the targeted nodes' positions. For the most part, each node requires a complete knowledge of the whole network topology. However, in distributed approaches, deployment decisions are made based solely on some local knowledge per node. For example, a deterministic deployment strategy for mobile data collecting nodes was proposed in Reference [26], assuming centralized knowledge and decisions made at the BS. These mobile nodes move along a set of pre-defined tracks in the sensing field. In the proposed deployment strat-egy, SNs were able to relay data in addition to their sensing duties. It was shown that using data collectors (mobile relays) extends the net-work lifetime compared to conventional WSNs using static SNs only. In fact, Data Collectors (DCs) were used earlier in References [27,31]. The network lifetime was divided into equal length time intervals called rounds. The DCs are relocated at the beginning of each round based on a centralized algorithm running at the BS. The objective was to minimize the aggregate consumed energy during one round. It was shown that the optimal locations, according to this objective function, remain optimal even when the objective becomes to min-imize the maximum energy consumed per SN. It should be noted that these two energy metrics are not suitable for finding the opti-mal locations of mobile nodes since the optimal solutions will not be functions of time, i.e., time independent. The reason is that the maxi-mum, or aggregate, energy consumed per round might not change with time; hence, locations of the data collector will not change with time. Consequently, the locations calculated may be far from opti-mal. Despite the advantages of these proposals, the deployed networks were prone to network partitioning and/or communication loss due to lack of fault-tolerance. In addition, these proposals are designed for a 2-D deployment problem, which is not the case in OEM applications. In OEM, a 3-D deployment plan for environment sensors is a must to achieve the desired outdoor observations [14,15]. Moreover, ignoring candidate sensory positions in the 3-D space can waste numerous opportunities in reducing energy consumptions based on closer locations and in achieving better connectivity performance. In Reference [28], a fault-tolerant random deployment was proposed. In particular, the authors proposed a distributed deployment algorithm to achieve a desired level of fault-tolerance for all SNs' WSN. The transmission power of every node is gradually increased until either the distance between two neighboring nodes exceeds a specific threshold or the maximum transmission power is reached. In this deployment, fault-tolerance is achieved at the expense of added cost. Transmission power adaptation requires complex hardware that raises the per node cost, hence increasing the overall network cost. In addition, power adaptation results in added energy consumption cause lifetime reduction. In Reference [29], another fault-tolerant WSN deployment was proposed. The authors considered the case where at least two disjointed paths exist between each pair of SNs. To achieve a desired level of fault-tolerance, deterministic RN placement was used. The problem was formulated as an optimization problem. However, it turned to be NP-hard. So a polynomial time approximating algorithm was proposed instead. This algorithm identifies candidates' positions for RNs that cover the maximum number of SNs while assuming a regular communication range shape. Thus, RNs' positions may not be accurate since it solely depends on the transmission ranges that are irregular in practice. Alternatively, fault-tolerance could be achieved through deploying spare (redundant) nodes. In fact, faulty RNs are even more harmful to the network than SNs. This behavior was shown in Reference [3] for a deterministic, grid-based deployment. Hence, an efficient fault-tolerance should account for faulty SNs as well as RNs.

7.1.1 Contributions

In this article, a comprehensive network deployment problem is considered based on decentralized algorithms running not only at the system sink (BS) but also at the core network nodes (SNs/RNs). SNs and RNs are jointly deployed. In addition, some MRNs are used to release the pressure from overloaded paths and fix any connectivity problems. Thus, the targeted problem can be stated as follows: *Given a 3-D deployment space and a limited number of SNs and static/mobile RNs, find the optimal positions that prolong the network lifetime* while maintaining connectivity and certain fault-tolerant constraints.* Accordingly, our main contributions towards solving this problem can be summarized as:

1. We overcome the huge search space of the candidate RNs' positions by finding a subset of the grid vertices for these RNs based on their intersecting communication ranges.
2. The optimization problem is divided into *initial deployment* and *periodic redeployment*. In the initial deployment, the optimal locations of all nodes are found. In the periodic deployment, MRNs are relocated based on a decentralized decision made by deployed nodes at their present positions.
3. Efficient energy metrics, the *minimum node residual energy* and the *total energy consumed*, are used to maximize the network lifetime. These two metrics guarantee an influential MRN relocation.
4. An upper bound for the maximum network lifetime in ideal operation conditions is derived. This bound is used to show the performance gains achievable by the proposed two-phase solution.

The remainder of this chapter is organized as follows. Section 7.2 describes the system model and the mathematical framework. The proposed deployment strategy is presented and discussed in Section 7.3. Section 7.4 presents the numerical results while conclusions are drawn in Section 7.5.

7.2 System Models

In this section we describe the communication model, the network architecture, and the lifetime model used in this article. All three models were tailored to suit OEM applications.

---

* See Definition 7.1.

7.2.1 Communication Model

In practice, the signal level at distance r from a transmitter varies depending on the surrounding environment. These variations are captured through the so called log-normal shadowing model. According to this model, the signal level at distance r from a transmitter follows a log-normal distribution centered on the average power value at that point [32]. Mathematically, this can be written as

$$P_d(d) = P_s - P_{loss}(d) = P_s - P_{loss}(r_0) - 10n \log\left(\frac{r}{r_0}\right) + \chi, \qquad (7.1)$$

where P_s is the transmission power, $P_{loss}(r_0)$ is the path loss measured at reference distance r_0 from the transmitter, n is an environment dependent path loss exponent, and χ is a normally distributed random variable with zero mean and variance σ^2, i.e., $\chi \sim N(0, \sigma^2)$. With the aid of this model, the probability of successful communication between two nodes separated with a distance r can be calculated as follows. Assume P_{min} is the minimum acceptable signal level for successful communication between a source S and a destination D separated by distance r. The probability of successful communication is $\rho[S,D] = P_r[P_d(r) \geq P_{min}]$. After some mathematical manipulations, $\rho[S,D]$ can be written as

$$\rho[S,D] = Q\left(\frac{P_{min} - P_s - P_{loss}(r_0) - 10n\log(r/r_0)}{\sigma}\right) \qquad (7.2)$$

where $Q(\cdot)$ is the Q-function defined as $Q(x) = \frac{1}{\sqrt{2\pi}} \int_x^\infty e^{-t^2/2} dt$. In this chapter, the probability of successful communication between nodes i and j should exceed a certain threshold, τ_1. Hence, the condition $\rho[i,j] \geq \tau_1$ will be used. The percentage of successful connectivity τ_1 is a design parameter. In fact, if identical transceiver specifications are used for all nodes, choosing a particular τ_1 automatically specifies the maximum distance between any two directly communicating nodes.

7.2.2 Network Model

Consider a WSN where nodes are logically grouped into two layers, a lower layer consisting of the SNs and an upper layer consisting of all the RNs. SNs forward their sensing data to a neighboring RN in the upper layer. On the other hand, RNs communicate periodically with the BS, either directly or via other RNs, to deliver the aggregated traffic from the SNs. Since these RNs have longer transmission range compared to SNs, SNs in such a 2-tier architecture can invest their energy only in data gathering. And RNs can take care of communicating the gathered data to the BS. This helps prolong the lifetime of the WSNs, which is highly desired in OEM applications operating in harsh and energy-constrained environments.

In the grid-based network architecture, the grid edge length is the transmission range r. Nodes can only be placed at the vertices of the 3-D grid such that the maximum number of Event Centers (ECs) is monitored. An EC is a location where the targeted phenomena can be monitored. The network topology is modeled as a graph $G = (V, E)$, where $V = \{n_0, n_1, \ldots, n_{v-1}\}$ is the set of v candidate grid vertices, E is the set of bidirectional links (edges) between the deployed nodes. Furthermore, the link between vertices i and j belongs to E if and only if the condition $\rho[i, j] \geq \tau_1$ is met. The network consists of Q_{SN} SNs and Q_{RN} RNs. If we let α_i^{SN} and α_i^{RN} be two binary variables such that $\alpha_i^{SN} = 1$ if an SN is placed at vertex i, $\alpha_i^{RN} = 1$ if an RN is placed at vertex i, and $\alpha_i^{SN} = 0$, $\alpha_i^{RN} = 0$ otherwise, then we can write

$$\sum_{i=1}^{v} \alpha_i^{SN} = Q_{SN}, \tag{7.3}$$

$$\sum_{i=1}^{v} \alpha_i^{RN} = Q_{RN}, \tag{7.4}$$

To achieve complete network coverage, the $Q_{SN} + Q_{RN}$ nodes should be distributed such that (1) every EC is covered by at least one SN, (2) every SN is connected to at least one RN, and (3) every RN is connected to the BS either directly or indirectly via other RNs. These three requirements can mathematically be written as follows. Let $\beta_{i,j}$ be a binary variable whose value is 1 if the ith vertex is a candidate position to sense the jth EC and 0 otherwise. Consequently, the first requirement can be written as

$$\sum_{i=1}^{v} \alpha_i^{SN} \cdot \beta_{ij} \geq 1, \quad \forall j \in S_{EC}, \tag{7.5}$$

where S_{EC} is the set of ECs. To guarantee the communication between the lower layer and the upper layer in the network, the second requirement can be written as

$$\sum_{i=1}^{v} \alpha_i^{SN} \cdot \alpha_j^{RN} \geq 1, \quad \forall i \in V \text{ and } j \in N(i) \qquad (7.6)$$

where $N(i)$ is a set of neighboring indices such that $j \in N(i)$ if the jth vertex is within the transmission range of the ith node, i.e., $\rho[i, j] \geq \tau_1$. Finally, to guarantee that every RN can reach the BS either directly (one hop) or indirectly (multiple hops), the third requirement can be written as

$$\alpha_j^{RN} \cdot \left(\sum_{i \in \{N(BS), M(N(BS))\}} \alpha_i^{RN} \right) \geq 1, \quad \forall j \notin N(BS), \qquad (7.7)$$

where $j \in M(N(BS))$ if the jth node can reach the BS either directly or indirectly.

7.2.3 Lifetime and Energy Models

Due to the harshness of outdoor environments, nodes and communication links are prone to failure. Losing some nodes and links may isolate other functional nodes. This problem can be overcome by deploying redundant nodes. Hence, deployment of redundant nodes helps achieving fault-tolerance, and thus prolongs the network lifetime. This concept can formally be defined as follows:

> **Definition 7.1:** (Network Lifetime): is the time span (in rounds) from network deployment to the instant when the percentage of alive* and connected irredundant SNs and RNs falls below a specific threshold τ_2.

Notice that the remaining nodes, in addition to being alive, need to be connected to the BS either directly or indirectly. In order to measure the network lifetime, a measuring unit needs to be defined. In this work, we adopt the concept of a round as the lifetime metric. A round is the time span t_{round} over which every EC reports to the BS at least once. At the end of every round, the total energy consumed by the ith node can be written as

---

* Alive nodes are those which have enough energy for at least one more round.

$$E_{cons}^i = \sum_{Per\ round} J_{tr} + \sum_{Per\ round} J_{rec}, \tag{7.8}$$

where $J_{tr} = L(\varepsilon_1 + \varepsilon_2 d^n)$ is the energy consumed for transmitting a data packet of length L to a receiver located r meters from the transmitter. Similarly, $J_{rec} = L\beta$ is the energy consumed for receiving a packet of the same length [33]. The parameters ε_1, ε_2, and β are hardware-specific parameters of the used transceivers. Accordingly, if the initial energy of the ith node, E_{init}^i, is known, its remaining energy, E_{rem}^i, at the end of the round can readily be calculated as $E_{rem}^i = E_{init}^i - E_{cons}^i$. At the end of every round, the total energy consumed by all nodes can be written as $E_{cons}^{tot} = \sum_{Q_{SN}} E_{cons}^i + \sum_{Q_{RN}} E_{cons}^i$. Since all RNs may transmit and/ or receive data in every round while all SNs transmit their measurements, the total energy consumed per round can be written as

$$E_{cons}^{tot} = \sum_{i=1}^{v} \alpha_i^{RN} \left(\sum_{j \in N(i)} J_{rec} f_{ij} + \sum_{j \in N(i)} J_{tr} f_{ij} \right) + \sum_{i=1}^{v} \alpha_i^{SN} \left(\sum_{j \in N(i)} J_{tr} f_{ij} \right),$$

$$\tag{7.9}$$

where f_{ij} is the traffic from node i to j measured in bits per second (bps). The way equations (7.3) through (7.9) are presented lends the network architecture and the energy model smoothly into the lifetime maximization problem discussed in the following section.

7.3 Deployment Strategy

The deployment problem studied in this chapter has an infinitely large search space. To limit this infinite search space to a manageable number of points, the 3-D grid model is used. The objective is to find the optimal locations of $Q_{SN} + Q_{RN}$ nodes among v grid vertices that maximize the network lifetime. The deployment strategy consists of two phases. The first phase aims to find the optimal positions of all the nodes such that total energy consumption is minimized. The second phase is launched at the end of every round to fix connectivity problem(s) and release the pressure from heavily loaded nodes. This two-phase deployment strategy is called *Optimized 3-D grid deployment (O3D)*.

7.3.1 First Phase of the O3D Strategy

Let us start with the *Simplest Deployment Scheme (SDS)*. SDS aims at maximizing the network lifetime by finding (1) the optimal deployment of all nodes in the network, and (2) the optimal routing paths from the SNs to the BS. This can be stated as follows: *what are the optimal deployment and routing strategies that need to be used to reduce the energy consumed per round?* While Equations (7.3) through (7.9) guarantee that a total of $Q_{SN} + Q_{RN}$ nodes satisfy the desired network topology, additional constraints are needed to control the

---

Algorithm 7.1: O3D first phase deployment

Function OptIniDep (*IS*: Initial Set of ECs and BS coordinates to construct N, V)

Input:

　A set *IS* of the ECs and BS coordinates.
　A set *V* of the candidate grid vertices.

Output:

　A set *N* of the SNs, SRNs, and BS coordinates.

Begin

　Initialize: k-values, v, S_{EC}, g_i^{SN}, g_i^{RN}, C_i^{SN}, C_i^{RN}, J_{tr}, J_{rec}, Q_{SN}, Q_{RN}.
　PS1 = Solve $\mathbb{P}_2$ in (7.18).
　N = Set of coordinates of SNs, SRNs, and BS in **PS1**.

End

---

routing paths. This can be done as follows. First, the traffic needs to be fairly divided among the deployed RNs to avoid node overload. Mathematically speaking, the conditions

$$\sum_{j\in N(i)} \alpha_i^{SN} \cdot f_{ij} \le g_i^{SN}, \quad \forall i \in V \tag{7.10}$$

$$\sum_{j\in N(i)} \alpha_i^{RN} \cdot f_{ij} - \sum_{k\in N(i)} \alpha_i^{RN} \cdot f_{ki} \le g_i^{RN}, \quad \forall i \in V \tag{7.11}$$

need to be met. g_i^{SN} and g_i^{RN} are the generated traffic from the ith SN and RN respectively, measured in bps. Secondly, the limit of the available bandwidth for every node needs to be maintained. This can be achieved through the conditions

$$\sum_{j\in N(i)} \alpha_i^{SN}\cdot f_{ij} \le C_i^{SN}, \qquad \forall i \in V \tag{7.12}$$

$$\sum_{j\in N(i)} \alpha_i^{RN}\cdot f_{ij} \le C_i^{RN}, \qquad \forall i \in V \tag{7.13}$$

where C_i^{SN} and C_i^{RN} are the available bandwidths for the ith SN and RN respectively, measured in bps. With the aid of these constraints, the SDS optimization problem can be summarized as follows $\mathbb{P}_1$:

$$\text{Minimize } E_{cons}^{tot}$$

$$\text{Subject to Equations } (7.3)-(7.9) \tag{7.14}$$

$$\text{Equations } (7.10)-(7.13)$$

Observe that we have intentionally divided the constraints into two groups, network architecture constraints and routing paths constraints. This division gives additional insights into the performance of the proposed scheme, which will be revealed in the results section. Despite that this scheme allows optimal deployment of a given set of nodes such that lifetime is maximized, the deployed nodes are prone to isolation and/or failure that render some ECs not covered. This is because the no fault-tolerance constraint was imposed. To achieve this tolerance, the *Fault-tolerant Simplest Deployment Scheme (FSDS)* presents a fault-tolerant version of the SDS. Fault-tolerance is an energy consuming constraint that allows the network to withstand a certain level of faulty nodes while maintaining a desired level of coverage. Fault-tolerance can be quantified by the number or percentage of faulty nodes tolerated. When the number of nonoperational nodes is $<k$, the network still needs to recover the isolated ECs and nodes. This is achieved through the injection of redundant nodes as mentioned earlier. Consequently, data recovery can be defined as:

Definition 7.2: (Data Recovery): The existence of operational redundant nodes capable of covering isolated or partitioned ECs and nodes and routing their measurements to the BS.

With the aid of these definitions, we can summarize the objective of the FSDS scheme as follows: maximize the network lifetime through finding (1) the optimal deployment of all nodes in the network, and (2) the optimal routing paths from the SNs to the BS, such that the deployed network is k fault-tolerant and data recovery is guaranteed. In other words, the FSDS scheme is an extension of the SDS with fault-tolerance and data recovery constraints. Let us start with the fault-tolerance constraint. Every component in the network (EC, SN, or RN) shall be connected to more than one element in the upper level to achieve a desired level of redundancy. In other words, every EC needs to be covered by $k_1 \geq 1$ SNs, every SN needs to be connected to at least $k_2 \geq 1$ RNs, and every RN needs to reach the BS through at least $k_3 \geq 1$ routes. Consequently, (7.5) through (7.7) can be rewritten as:

$$\sum_{i=1}^{v} \alpha_i^{SN} \cdot \beta_{ij} \geq k_1, \qquad \forall j \in S_{EC} \tag{7.15}$$

$$\sum_{i=1}^{v} \alpha_i^{SN} \cdot \alpha_j^{RN} \geq k_2, \qquad \forall j \in V \text{ and } j \in N(i), \tag{7.16}$$

$$\alpha_j^{RN} \cdot \left(\sum_{i \in \{N(BS), M(N(BS))\}} \alpha_i^{RN} \right) \geq k_3, \quad \forall j \notin N(BS) \tag{7.17}$$

It should be mentioned here that the way Equations (7.15) through (7.17) are written gives us flexibility in choosing the level of tolerance at all layers in the network. This allows more customized fault-tolerance. However, to make the entire network a k fault-tolerant network, we simply set $k_1 = k_2 = k_3 = k$. Finally, $\mathbb{P}_1$ can be rewritten with the modified constraints to get the FSDS in the form of $\mathbb{P}_2$:

$$\text{Minimize} \quad E_{cons}^{tot}$$

$$\text{Subject to} \quad \text{Equations } (7.3)-(7.9),$$

$$\text{Equations } (7.10)-(7.13), \tag{7.18}$$

$$\text{Equations } (7.15)-(7.17)$$

By solving this optimization problem at the system sink (BS), a fault-tolerant, maximized lifetime WSN deployment will be achieved, which is also the first phase deployment of the O3D strategy summarized in Algorithm 7.1. An additional degree of freedom can be brought to the network through the use of MRNs. MRNs can be reallocated periodically such that a particular objective is achieved. In our case, we shall use it to maximize the network lifetime by reducing the energy consumption of a particular set of nodes that have been overloaded as described next.

7.3.2 Second Phase of the O3D Strategy

After performing the first phase of the O3D in $\mathbb{P}_2$ which considers all the available RNs, i.e., Q_{RN}^S static RNs and Q_{RN}^M mobile RNs, $\left(Q_{RN} = Q_{RN}^S + Q_{RN}^M\right)$, a relocation of the MRNs is performed in the second phase of the O3D strategy, which will subsequently be shown. However, since the second phase deployment will take place during the network run time, processing time is very critical and has to be very limited. Thus, searching all the grid vertices, v, in large-scale applications with large v like OEM applications, is a computationally expensive and time consuming process. This is due to the involved computational requirements for finding the network lifetime

for a large number, $\left(\dfrac{v - Q_{RN}^S - Q_{SN}}{Q_{RN}^M} \right)$, of possible node locations.

Therefore, a limited search space with reduced v is needed. Taking advantage of a distributed system running at the SNs and the static RNs to provide a previous knowledge of their positions to the BS, the MRNs may be placed on any grid vertex as long as they are within the probabilistic communication range of the largest number of SNs and/or static RNs. As a result, the search space is reduced while the accuracy of the deployment plan is not affected. To explain our method of finding this finite search space, we give the following two definitions:

Definition 7.3: (Ideal Set): A finite set of positions P is ideal if and only if it satisfies the following property: there exists an optimal placement of MRNs in which each relay is placed at a position in P.

We try to find such an ideal set in order to achieve a more efficient, discrete search space in which candidate positions do not include all the grid vertices but only a subset of them. This subset should have the highest potential to prolong the network lifetime and sustain its fault-tolerance through maintaining the largest neighborhood. However, since computational complexity is proportional to the cardinality of P, a set with an even smaller cardinality is needed.

Definition 7.4: (Candidate Grid Unit (CnGU)): A candidate grid unit α is a grid unit that has a connected center with at least k_2 SN or k_3 static RN. The subset of SNs and static RNs coordinates connected to α is denoted by $C(\alpha)$.

The building unit of the 3-D grid is called a *Grid Unit (GU)*. A *GU* is said to be connected to a particular SN or static RN if the condition $\rho \geq \tau_1$ is met, where ρ is the probabilistic connectivity parameter between the *GU* center and that node.

Definition 7.5: (Optimal Candidate Grid Unit (OCGU)): A candidate GU α is optimal if there is no candidate GU β, where $C(\alpha) \subseteq C(\beta)$.

The OCGUs have the highest potential to place the MRNs based on Definitions 7.4 and 7.5. Accordingly; we have to show that an ideal set can be derived from the set of OCGUs. Towards this end, we state the following *Lemmas*:

Lemma 7.1: For every CnGU β, there exist a OCGU α such that $C(\beta) \subseteq C(\alpha)$.

Proof: If β is an OCGU, we choose α to be β itself. If β is not an OCGU then, by definition, there exists a CnGU α_1 such that $C(\beta) \subseteq C(\alpha_1)$. If α_1 is an OCGU, we choose β to be α_1, and if α_1 is not an OCGU then, by definition, there exists another CnGU α_2 such that $C(\alpha_1) \subseteq C(\alpha_2)$. This process continues until an OCGU α_x is found; we choose α to be α_x. Thus, Lemma 7.1 holds. ∎

Lemma 7.2: Finding an OCGU takes at most $(n-1)$ step, where $n = Q_{SN} + Q_{RN}^S$.

Proof: By referring to the proof of Lemma 7.1, it is clear that $|C(\alpha_x)| \leq n$, and $|C(\alpha)| \langle |C(\alpha_1)| \langle |C(\alpha_2)| \langle ... \langle |C(\alpha_x)| \leq n$;

where $|C|$ is the cardinality of C. Consequently, the process of finding the OCGU α_x takes a finite number of steps $\leq n-1$.

Then, we introduce the following *Theorem*:

Theorem 7.1: A set P that contains one position from every OCGU is ideal.

Proof: To prove this Theorem, it is sufficient to show that for any arbitrary placement Z we can construct an equivalent[*] placement $\bar{Z}$ in which every MRN is placed at a position in P. To do so, assume that in Z, an MRN i is placed such that it is connected to a subset J of SNs/SRNs. It is obvious that there exists a CnGU β, such that $J \subseteq C(\beta)$. From Lemma 7.1, there exist an OCGU α such that $C(\beta) \subseteq C(\alpha)$. In $\bar{Z}$, we place i at the position in P that belongs to α, so that i is placed at a position in P and is still connected with all SNs/SRNs in J. By repeating for all MRNs, we construct a placement $\bar{Z}$ which is equivalent to Z, and thus Theorem 7.1 holds.   ∎

In order to find all OCGUs, we need a data structure associated with each GU to store coordinates and total number of SNs/SRNs connected to the GU. We represent this data structure by the CnGU set $C(i)$, where i is the center of the CnGU. By computing $C(i)$, $\forall\ i \in V$, we can test whether a CnGU centered at i is optimal or not by searching for a set that has at least all elements of $C(i)$. In the following, Algorithm 7.2 is running independently at the SNs/RNs by the end of each round to collect residual energy and neighboring CnGUs per node. If any change occurs in the resultant gathered information, it will be broadcasted to update the system sink. Accordingly, Algorithms 7.3 through 7.6 will be executed at the system sink (BS). Algorithm 7.3 establishes the set of vertices that are within the communication range of a single GU based on the output of Algorithm 7.2; it runs in $O(n)$ time. Algorithm 7.4 tests whether a CnGU is optimal or not based on a local comparison between the resultant output of each SN/RN.

---

[*] Equivalent in terms of connected SNs/SRNs. In other words, the placement of an MRN at position i, within the communication range of the nodes x and y, is equivalent to the placement of the same MRN node at position j within the communication range of the nodes x, y, and z.

Algorithm 7.5 uses Algorithms 7.3 and 7.4 to construct the ideal set P by finding all OCGUs. The overall complexity of Algorithm 7.5 is $O(n \log n)$.

Once we obtain the set P which contains one position (grid vertex coordinates) from each OCGU, the search space of the optimization problem to be formulated, subsequently, becomes much more limited. To start formulating the second phase optimization problem, we assume that an FSDS deployment has already been performed and that the network is running. Some nodes will be consuming more energy than others such that network bottlenecks start to appear.

Algorithm 7.2: Information gathering

Function InfoG(N)

Input:

A set N of the GUs' coordinates.

Output:

LoCGU: List of covered GUs by an SN/SRN j.
E: remaining energy at SN/SRN j.

Begin

If SN

$E = E_{rem}^{SN}$ **with reference to Equation (7.9)**

Endif

If RN
$E = E_{rem}^{RN}$ **with reference to Equation (7.9)**

Endif

LoCGU$(j) := \varnothing$; //list of covered GUs by node j

foreach GU center i **do**

Compute $\rho[i, j]$;

If $\rho[i, j] \geq \tau_1$

LoCGU$(j) := i \cup$ *LoCGU*(j);

endif

endfor

End

Algorithm 7.3: Creating grid units

Function FindCandidateGridUnit(*N*, *LoCGUs*)

Input:

N: set of the SNs' and SRNs' coordinates.
LoCGUs: list of covered GUs by each SN/SRN.

Begin

foreach GU center *i* in *LoCGUs* **do**

$C(i) := \varnothing$;

$x_i := 0$; //where x_i represents number of nodes connected to the GU centered at *i*.//

$L_i := \varnothing$; //L_i is the list of SN/SRN coordinates connected to the GU center *i*.//

foreach SN/SRN *j* in *LoCGUs* do

If *i* is covered by an SN/SRN
$x_i = x_i + 1$;
$L_i = L_i \cup j$;

endif

endfor

If$(x_i \geq k)$
$C(i) := L_i \cup C(i)$;

Endif

endfor

End

Algorithm 7.4: Testing whether $C(i)$ is Optimal or Not

Function Optimal($C(i)$, all non-empty grid unit sets)

Input:

A set $C(i)$ for a specific grid unit center i and All non-empty sets of the grid units' centers.

Output:

True if $C(i)$ is an OCGU and False otherwise.

Begin

If $C(i) := \varnothing$ **do**

return False;

endif

Search for a set $\bar{C}$ such that $C(i) \subseteq \bar{C}$.

If $\bar{C} := \varnothing$ **do**

return True;

else

return False;

endif

End

Algorithm 7.5: Finding all OCGUs

Function FindOCGUs(N: Set of SNs and SRNs)

Input:

A set N of the SNs' and SRNs' coordinates.

Output:

A set P that contains one position from every OCGU.

Begin

$P := \emptyset$;

FindCandidateGridUnit (N);

foreach $C(i)$ do

    If Optimal($C(i)$, all non-empty grid unit sets) **do**
    $P := \{i\} \cup P$;

    endif

endfor

End

Algorithm 7.6: O3D second phase deployment

Function MRNsP(N: constructed by SNs, SRNs, and BS, P)

Input:

    A set N of the SNs, SRNs, and BS nodes' coordinates.
    An ideal set $PS2$ of v candidate positions for the MRNs.

Output:

    A set PS2 of the MRNs coordinates maximizing lifetime of
    N.

Begin

    Initialize: k-values, S_{EC}, g_i^{SN}, g_i^{RN}, C_i^{SN}, C_i^{RN}, Q_{RN}^{M}, And
    make $V = P$, $v = |P|$.
    $PS2 = $ Solve $\mathbb{P}_3$ in (7.23).

End

To help these nodes, a total of Q_{RN}^{M} MRNs shall be relocated. The objective of this relocation process is twofold: maximize the minimum residual energy among all nodes per round and minimize the total energy consumption. Notice that the first part of the objective aims at maximizing a time- and node-dependent value; hence, relocation will always lead to new optimal locations whenever applied. The objective function can be written as

$$\text{Minimize} \quad E_{cons}^{tot} - E_{res}^{SN} - E_{res}^{RN} \tag{7.19}$$

where E_{res}^{SN} and E_{res}^{RN} are the minimum residual energy over all SNs and RNs respectively, at the end of the round. Notice that RNs include both static RNs and MRNs. To guarantee connectivity for the entire round, these two residual energies need to satisfy two conditions at the end of the round. The first one is

$$E_{res}^{SN}, E_{res}^{RN} \geq 0 \tag{7.20}$$

while the second one is

$$\alpha_i^{SN} E_{rem}^{SN} - \sum_{j \in N(i)} \alpha_i^{SN} J_{tr} f_{ij} - \sum_{k \in N(i)} \alpha_i^{SN} J_{rec} f_{ki} \geq E_{res}^{SN}, \quad \forall i \in V$$

$$\alpha_i^{RN} E_{rem}^{RN} - \sum_{j \in N(i)} \alpha_i^{RN} J_{tr} f_{ij} - \sum_{k \in N(i)} \alpha_i^{RN} J_{rec} f_{ki} \geq E_{res}^{RN}, \quad \forall i \in V \tag{7.21}$$

where E_{rem}^{SN} and E_{rem}^{RN} are the remaining energies in the SNs and RNs respectively. These two equations guarantee successful communication for at least one coming round. Furthermore, at the end of the relocation process, the number of MRNs needs be the same, whether relocated or kept in place. Hence, similar to Equations 7.3 and 7.4, one can write

$$\sum_{i=1}^{v} \alpha_i^{MRN} = Q_{RN}^{M}. \tag{7.22}$$

Consequently, with the aid of these constraints, we can rephrase our optimization to be written as $\mathbb{P}_3$:

$$\text{Minimize} \quad E_{cons}^{tot} - E_{res}^{SN} - E_{res}^{RN},$$

$$\text{Subject to Equations } 7.3 - 7.9,$$

$$\text{Equations } 7.10 - 7.13, \tag{7.23}$$

$$\text{Equations } 7.15 - 7.17,$$

$$\text{Equations } 7.20 - 7.22$$

By solving this optimization problem, a further fault-tolerant and maximized lifetime WSN deployment will be achieved; this is also the second phase deployment of our O3D strategy summarized in

Algorithm 7.6. In light of the large existing literature today on optimization formulations for lifetime in WSNs, it is worth pointing out that the two energy metrics (residual and consumed energy metrics) included in the objective function make the proposed Mixed Integer Linear Program (MILP) unique and particularly distinguished. As it makes the proposed approach more suitable for finding the optimal locations of mobile nodes since the optimal solutions will be a function of time, i.e., time-dependent. This is a typical use of the proposed mathematical model under realistic settings in mobile topologies.

Finally, based on the output of Algorithms 7.1 and 7.6, optimal locations of SNs and RNs in terms of maximum lifetime and limited cost budget are determined as are fault-tolerance and data recovery constraints. This two-phase solution can easily be extended to consider other constraints such as coverage, data fidelity, and delay tolerance through formulating and adding it to $\mathbb{P}_3$.

7.4 Lifetime Theoretical Analysis

In the previous section, we examined the placement problem for a heterogeneous WSN when both energy efficient and fault-tolerant design factors are considered. However, once the network is operational, deployed nodes start losing energy and face harsh operational conditions that may lead to failures and increased risks of disconnection. Consequently, a second phase was proposed to reposition MRNs in order to overcome such conditions, and hence provide near optimal solutions in practical situations. As these conditions are scarcely predictable in practice, it is very difficult to predict what would be the maximum number of rounds for which a network can stay operational. Therefore, we derive an Upper Bound (UB) on the number of rounds a WSN can spend during its lifetime, given that there are no unexpected node/link failures. Thereby, we assume the same notations in the system models section. Also, we define LT_{max} to be the maximum number of rounds a WSN can stay operational for, and $E_{min/r}^{SN}$ to be the minimum total energy consumed by SNs per round, and $E_{min/r}^{RN}$ to be the minimum total energy consumed by RNs per round. Assume E_{init}^{tot} is the initial total available energy before the network starts functioning, E_{init}^{SN} is the initial available energy per SN, and E_{init}^{RN} is the initial available energy per RN.

Theorem 7.2: The lifetime of the deployed WSN is upper bounded by:

$$
LT_{max} = \min \left\{ \frac{Q_{SN} \cdot E_{init}^{SN}}{J_{tr} \left[\dfrac{Q_{SN}}{k_1} \displaystyle\sum_{i=1}^{\frac{Q_{SN}}{k_1}} g_i^{SN} \right]}, \frac{Q_{RN} \cdot E_{init}^{RN}}{\left(J_{tr} + J_{rec} \right) \left[\dfrac{Q_{RN}}{k_2} \displaystyle\sum_{i=1}^{\frac{Q_{RN}}{k_2}} g_i^{RN} \right]} \right\}
$$

$$(7.24)$$

Proof: As the minimum consumed energy per round by SNs is the required energy to deliver irredundant generated traffic (sensed data), the minimum energy consumed by these nodes per round is equal to the energy used in transmitting from irredundant SNs. Since the number of irredundant SNs is Q_{SN}/k_1; the minimum energy consumed per round is

$$
E_{min/r}^{SN} = J_{tr} \left[\frac{Q_{SN}}{k_1} \sum_{i=1}^{\frac{Q_{SN}}{k_1}} g_i^{SN} \right] \qquad (7.25)
$$

Similarly, the number of irredundant RNs is Q_{RN}/k_2; hence, the minimum energy consumed per round is

$$
E_{min/r}^{RN} = \left(J_{tr} + J_{rec} \right) \left[\frac{Q_{RN}}{k_2} \sum_{i=1}^{\frac{Q_{RN}}{k_2}} g_i^{RN} \right] \qquad (7.26)
$$

As the initial total available energy at SNs is equal to $Q_{SN} \cdot E_{init}^{SN}$, the maximum number of rounds the SNs can stay operational for is $Q_{SN} \cdot E_{init}^{SN} / E_{min/r}^{SN}$. Similarly, the maximum number of rounds the RNs can stay operational for is $Q_{RN} \cdot E_{init}^{RN} / E_{min/r}^{RN}$. As the maximum number of rounds a WSN can stay operational for is controlled by the lifetime of SNs generating the sensed data, and the RNs relaying this data, the maximum number of rounds a WSN can stay operational for is

$$LT_{max} = min\left\{\frac{Q_{SN} \cdot E_{init}^{SN}}{E_{min/r}^{SN}}, \frac{Q_{RN} \cdot E_{init}^{RN}}{E_{min/r}^{RN}}\right\} \qquad (7.27)$$

By substituting Equations (7.25) and (7.26) in Equation (7.27), we achieve the lifetime UB described above in Equation (7.24).

■

It is worth mentioning that this UB depends only on the number of deployed nodes, initial node energy, node generation rate, redundancy level (i.e., k-value), and energy consumed for transmitting/receiving a packet. Consequently, to increase LT_{max} one can either increase the initial energy of the deployed nodes, decrease their energy consumption per packet, increase the number of nodes deployed, or increase the redundancy level. This UB is not only used in assessing the efficiency of our two-phase deployment strategy but also in any other deployment strategy aiming at maximizing the WSN lifetime.

7.5 Performance Evaluation and Discussion

In this section, we evaluate the performance of our proposed strategy in practical settings with different PNF and PDN conditions. We consider the SDSs and FSDSs along with the UB as a baseline to the proposed O3D deployment strategy. In fact, simplified variations of SDSs and FSDSs are widely studied in the literature [27,31]. To compare the performance of the three schemes, the following four performance metrics are used. The first metric is the average lifetime defined as the number of rounds through which the network operates. The second one is the average energy consumed per byte. This metric reflects the energy utilization efficiency. In fact, it relates energy consumption to network lifetime. The third metric is the *Ratio of Remaining Energy (RRE)*. RRE is the ratio of the total remaining energy in all nodes to their total initial energy. Finally, the fourth metric is the *Percentage of Packet Loss (PPL)*. PPL is the percentage of transmitted data packets that fails to reach the BS. It reflects the effects of bad communication channels and node failures. In studying these performance metrics, four parameters are used. These are the level of fault-tolerance k, the PNF, the PDN, and the $(Q_{SN} + Q_{RN})$ count.

7.5.1 Simulation Model

The three deployment schemes, SDS, FSDS, and O3D, are applied based on experimental and realistic parameters that are typical in outdoor WSN applications [32,34]. Moreover, these schemes are executed on 500 randomly generated WSNs' hierarchical graph topologies in order to get statistically stable results. The average results hold confidence intervals of no more than 2% of the average values at a 95% confidence level. For each topology, we apply a random node/link failure based on the pre-specified PNF and PDN values, and performance metrics are computed accordingly. Dimensions of the monitored space are $900 \times 900 \times 300 \, m^3$. We assume the same predefined fixed time schedule in Reference [18] for traffic generation at the SNs and RNs, as it is a typical model for such hierarchical topologies in WSNs. Nodes' positions are found by applying the three deployment strategies. We assume that each WSN is required to be operational for the maximum number of rounds using a maximum of 15 RNs and 1500 SNs (cost constraint). Based on real outdoor experimental measurements, the communication model parameters are set as shown in Table 7.2 [34]. The simulator determines whether or not an SN is connected to its neighbors according to the probabilistic communication model described earlier in Section 2.1. To simplify the presentation of the results, all the transmission ranges of SNs and RNs are assumed equal to 100 m. We use MATLAB® *lp-solver v5.5* with a timeout of 15 min. In other words, the MILP of a particular round is solved during the last 15 min of the previous round. We observe that better lifetime performance can be achieved at the cost of more computational time complexity. However, this is not an issue in a typical OEM application, where an MILP can be left running for days without affecting the targeted application.

7.5.2 Simulation Results

While the SDS strategy finds the optimal locations of the SNs and RNs in order to achieve the maximum network lifetime with cost constraints, FSDS and O3D attempt to find locations of these nodes such that the network lifetime is maximized under certain cost and fault-tolerance constraints. It is expected that the system lifetime

Table 7.2   Parameters of the Simulated WSNs

PARAMETER	VALUE	PARAMETER	VALUE	PARAMETER	VALUE	PARAMETER	VALUE
τ_1	70%	L	512 (bits)	δ^2	10	r	100 (m)
Q_{SN}	1500	E_{init}^{SN}	3000 (J)	Q_{RN}^{M}	5	PDN	20%
ε_1	50×10^{-9} (J/bit)	V	300	k	3	Q_{RN}	15
ε_2	10×10^{-12} (J/bit/m^2)	P_{min}	-104 (dB)	g_i^{SN}	10 (byte/round)	C_i^{SW}	1000 (byte/h)
β	50×10^{-9} (J/bit)	t_{round}	24 (h)	g_i^{RN}	100 (byte/round)	C_i^{RN}	2000 (byte/h)
n	4.8	PNF	20%	τ_2	70%	E_{init}^{RN}	3000 (J)

(measured in rounds) improves as Q_{SN} and Q_{RN} increased in a given terrain. This behavior is depicted in Figure 7.1 below. The more active nodes available in a given terrain the better the connectivity is and the lesser partitions are formed in the network. Consequently, this prolongs the network lifetime as mentioned earlier. The major effect of fault-tolerance constraints obviously appears in practical situations, where PNF and PDN are nonzero as depicted in Figures 7.2 and 7.3. According to results obtained by applying Theorem 7.2, the mobility factor has a great influence on the efficiency of the proposed two-phase deployment scheme by which we can achieve an average of 97% of the UB. The small difference between results obtained by O3D and UB is due to the assumed probabilistic communication model that causes additional transmitting and receiving processes when bad channel conditions are experienced. Moreover, comparing the number of rounds achieved by the SDS and the FSDS strategies with those achieved by the UB clearly shows the performance gains achieved by the O3D scheme over the two.

It is shown in Figure 7.1 that the SDS and FSDS were only able to achieve an average of 50% of the UB values. Figure 7.2 compares the PNF with respect to average lifetime measured in rounds assuming

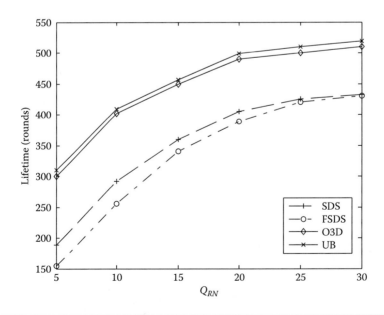

Figure 7.1　Lifetime as a function of the number of RNs.

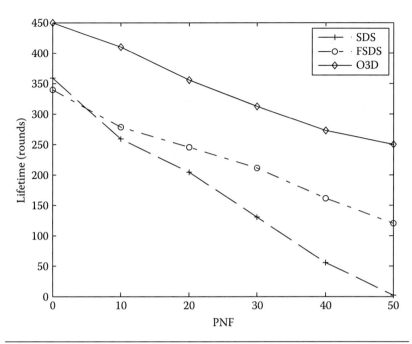

Figure 7.2   Lifetime as a function of the Probability of Node Failure (PNF).

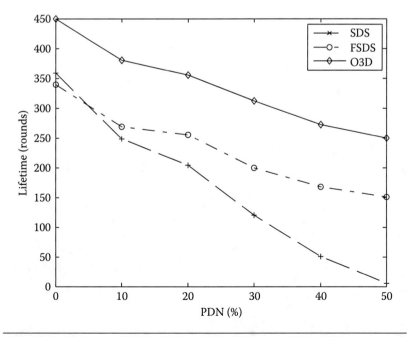

Figure 7.3   Lifetime as a function of the Probability of Disconnected Nodes (PDN).

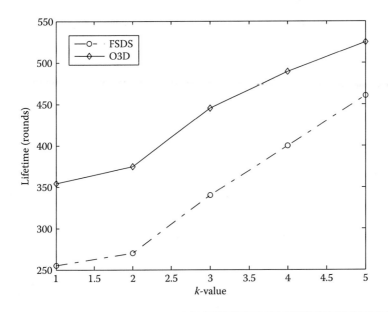

Figure 7.4 Lifetime as a function of fault-tolerance level k.

$Q_{RN} = 15$ and $Q_{SN} = 100 \times Q_{RN}$. Apparently, FSDS achieves better results than SDS. This is also the case when the PDN is compared against average lifetime in Figure 7.3. Figure 7.4 compares the FSDS and the O3D in terms of the network lifetime under different fault-tolerance levels. The PDN and PNF are set to 20%. It can be seen that larger k values, about 80% more, are needed for the FSDS to give the same lifetime as the O3D. This feature is appealing in situations where node mobility is either too expensive or even impossible. A similar conclusion can be drawn from Figure 7.5 for the PPL performance. Figure 7.6 shows the monotonically increasing relation between the average energy consumed and the PNF. It can be seen that the higher the percentage of faulty nodes the more energy is consumed. This causes poor connectivity as well as extended communication delays, which are undesirable properties in OEM applications. However, combining the mobility feature with the fault-tolerance constraint, in O3D, leads to stable energy consumption per byte under varying PNF values. Figure 7.7 illustrates the dependence between the RRE and the PDN. It shows that the higher the PDN, the energy is left unused, i.e., the energy remained in the nodes while the network is deemed dead. This causes degraded connectivity which leads to more

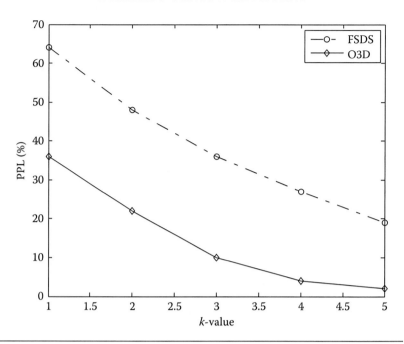

Figure 7.5   Packet loss as a function of the fault-tolerance level k.

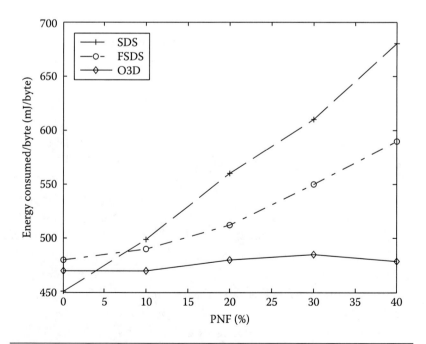

Figure 7.6   Energy consumed vs. the Probability of Node Failures (PNFs).

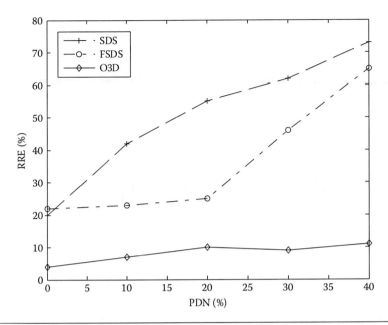

Figure 7.7 Ratio of Remaining Energy (RRE) vs. PDN.

partitions, thus terminating the network's lifetime. Figure 7.7 also shows how the RRE remains stable as long as the network has enough of a fault-tolerant level. For example, the RRE of the networks generated by the FSDS remains around 23% as long as the PDN is ≤20%. For the PDN >20%, the RRE rapidly increases as the PDN increase. However, this problem does not appear with the O3D strategy due to its ability to replace the deployed nodes at the beginning of each round. Choosing an appropriate k-value highly depends on the PNF and the PDN as illustrated in Figure 7.8. For a low PNF, a low redundancy level is needed. On the other hand, k must be at least equal to 4 to guarantee the network functionality for at least 300 rounds in environments with a 50% PNF. Similar results can be obtained to determine the choice of k under specific PDN values.

7.6 Conclusions

This chapter proposed a jointly energy-efficient and $k-$ fault-tolerant node deployment strategy for heterogeneous WSNs. Intensive simulations showed that jointly considering energy efficiency and fault-tolerance in node deployment can dramatically increase the network

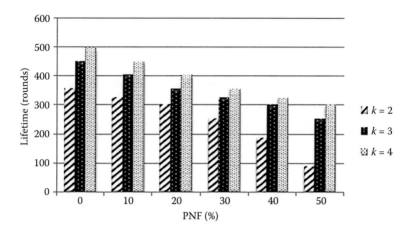

Figure 7.8   Lifetime vs. the Probability of Node Failure under three different fault-tolerance levels; $k = 2, 3$, and 4.

lifetime in OEM applications. To maintain these two objectives during the operation time, a certain number of MRNs is used. These MRNs can be relocated based on decentralized decisions made at certain points in time such that energy efficiency and fault-tolerance are maintained. To find the optimal positions of these MRNs, an optimization problem is formulated where the objective is to jointly minimize the total energy consumed and maximize the minimum residual energy. It was shown that the proposed combined scheme with fault-tolerance, energy efficiency, and MRNs outperforms the previously proposed schemes in withstanding the harsh OEM conditions which cause both node and link failures. We evaluated the proposed scheme extensively based on a real outdoor dataset and presented a distributed implementation of WSN in OEM. This implementation introduces efficient load balanced reporting. It shows that an improvement of 40% in the network lifetime can be achieved in comparison to classical approaches in the literature where the three factors, *fault-tolerance*, *energy efficiency*, and *MRNs*, are ignored.

References

1. M. Younis and K. Akkaya, Strategies and techniques for node placement in wireless sensor networks: A survey, *Elsevier Ad Hoc Networks Journal*, vol. 6, no. 4, pp. 621–655, 2008.

2. I. Akyildiz, W. Su, Y. Sankarasubramaniam, and E. Cayirci, A survey on sensor networks, *IEEE Communications Magazine*, vol. 40, no. 8, pp. 102–114, 2002.
3. F. Al-Turjman, H. Hassanein, and M. Ibnkahla, Quantifying connectivity in wireless sensor networks with grid-based deployments, *Elsevier Journal of Network and Computer Applications*, vol. 36, no. 1, pp. 368–377, 2013.
4. P. Wang, Z. Sun, M. Vuran, M. Al-Rodhaan, A. Al-Dhelaan, and I. Akyildiz, On network connectivity of wireless sensor networks for sandstorm monitoring, *Elsevier Computer Networks Journal*, vol. 55, no. 5, pp. 1150–1157, 2011.
5. I. Akyildiz and E. Stuntebeck, Wireless underground sensor networks: Research challenges, *Elsevier Ad Hoc Networks Journal*, vol. 4, pp. 669–686, 2006.
6. I. Akyildiz, D. Pompili, and T. Melodia, Underwater acoustic sensor networks: Research challenges, *Elsevier Ad Hoc Networks Journal*, vol. 3, pp. 257–279, 2005.
7. M. Hashim and S. Stavrou, Measurements and modeling of wind influence on radiowave propagation through vegetation, *IEEE Transactions on Wireless Communications Journal*, vol. 5, no. 5, pp. 1055–1064, 2006.
8. K. Akkaya and F. Senel, Detecting and connecting disjoint sub-networks in wireless sensor and actor networks, *Elsevier Ad Hoc Networks Journal*, vol. 7, no. 7, pp. 1330–1346, 2009.
9. I. Stojmenović, D. Simplot-Ryl, and A. Nayak, Toward scalable cut vertex and link detection with applications in wireless ad hoc networks, *IEEE Network*, vol. 25, no. 1, pp. 44–48, 2011.
10. F. Wang, M. Thai, and D. Du, On the construction of 2-connected virtual backbone in wireless networks, *IEEE Transactions on Wireless Communications Journal*, vol. 8, no. 3, pp. 1230–1237, 2009.
11. F. Al-Turjman, H. Hassanein, and M. Ibnkahla, Connectivity optimization with realistic lifetime constraints for node placement in environmental monitoring, In *Proceedings of IEEE Conference on Local Computer Networks (LCN)*, Zürich, Switzerland, 2009, pp. 617–624.
12. A. Erman, L. Hoesel, P. Havinga, and J. Wu, Enabling mobility in heterogeneous wireless sensor networks cooperating with UAVs for mission-critical management, *IEEE Transactions on Wireless Communications Journal*, vol. 15, no. 6, pp. 38–46, 2008.
13. P. Bellavista, A. Corradi, and C. Giannelli, Mobility-aware middleware for self-organizing heterogeneous networks with multihop multipath connectivity, *IEEE Transactions on Wireless Communications Journal*, vol. 15, no. 6, pp. 22–30, 2009.
14. D. Pompili, T. Melodia, and I. Akyildiz, Three-dimensional and two-dimensional deployment analysis for underwater acoustic sensor networks, *Elsevier Ad Hoc Networks Journal*, vol. 7, no. 4, pp. 778–790, 2009.
15. G. Tolle, J. Polastre, R. Szewczyk, and D. Culler, A microscope in the redwoods, In *Proceedings of ACM Conference on Embedded Networked Sensor Systems*, San Diego, CA, 2005, pp. 51–63.

16. B. Son, Y. Her, and J. Kim, A design and implementation of forest-fires surveillance system based on wireless sensor networks for South Korea mountains, *International Journal of Computer Science and Network Security*, vol. 6, no. 9, pp. 124–130, 2006.

17. F. Al-Turjman, H. Hassanein, and M. Ibnkahla, Efficient deployment of wireless sensor networks targeting environment monitoring applications, *Elsevier Computer Communications Journal*, vol. 36, no. 2, pp. 135–148, 2013.

18. F. Al-Turjman, H. Hassanein, W. Alsalih, and M. Ibnkahla, Optimized relay placement for wireless sensor networks federation in environmental applications, *Wireless Communication and Mobile Computing, Wiley*, vol. 11, no. 12, pp. 1677–1688, 2011.

19. K. Akkaya and M. Younis, A survey on routing protocols for wireless sensor networks, *Journal of Ad Hoc Networks*, vol. 3, no. 3, pp. 325–349, 2005.

20. H. O. Tan and I. Körpeoğlu, Power efficient data gathering and aggregation in wireless sensor networks, ACM SIGMOD Record, v.32 n.4, pp. 66–71, 2003.

21. H. O. Tan, I. Körpeoğlu, and I. Stojmenović, Computing localized power-efficient data aggregation trees for sensor networks, *IEEE Transactions on Parallel and Distributed Systems*, vol. 22, no. 3, pp. 489–500, 2011.

22. S. Hussain and O. Islam, An energy efficient spanning tree based multi-hop routing in wireless sensor networks, *Wireless Communications and Networking Conference (WCNC), IEEE*, Kowloon, China, 2007, pp. 4383–4388.

23. K. Kalpakis, K. Dasgupta, and P. Namjoshi, Maximum lifetime data gathering and aggregation in wireless sensor networks, In *Proceedings of the 2002 IEEE International Conference on Networking (ICN'02)*, 2002, pp. 685–696.

24. K. Xu, H. Hassanein, G. Takahara, and Q. Wang, Relay node deployment strategies in heterogeneous wireless sensor networks, *IEEE Transactions on Mobile Computing*, vol. 9, no. 2, pp. 145–159, 2010.

25. H. Tan, I. Körpeoğlu, and I. Stojmenović, Computing localized power efficient data aggregation trees for sensor networks, *IEEE Transactions on Parallel and Distributed Systems*, vol. 22, no. 3, pp. 489–500, 2011.

26. W. Alsalih, H. Hassanein, and S. Akl, Routing to a mobile data collector on a predefined trajectory, In *Proceedings of IEEE International Conference on Communications (ICC)*, Dresden, Germany, 2009, pp. 1–5.

27. M. Dawande, R. Prakash, S. Venkatesan, and S. Gandham, Energy efficient schemes for wireless sensor networks with multiple mobile base stations, In *Proceedings of IEEE Global Telecommunications Conference (GLOBECOM)*, San Francisco, CA, 2003, pp. 377–381.

28. M. Ishizuka and M. Aida, Performance study of node placement in sensor networks, In *Proceedings of International Conference on Distributed Computing Systems Workshops (ICDCS)*, Tokyo, Japan, 2004, pp. 598–603.

29. B. Hao, H. Tang, and G. Xue, Fault-tolerant relay node placement in wireless sensor networks: Formulation and approximation, In *Proceedings of Workshop on High Performance Switching and Routing (HPSR)*, Phoenix, AZ, 2004, pp. 246–250.
30. K. Akkaya, M. Younis, and M. Bangad, Sink repositioning for enhanced performance in wireless sensor networks, *Computer Networks*, vol. 49, no. 4, pp. 512–534, 2005.
31. A. Azad and A. Chockalingam, Mobile base stations placement and energy aware routing in wireless sensor networks, In *Proceedings of IEEE Wireless Communications and Networking Conference (WCNC)*, Las Vegas, NV, 2006, pp. 264–269.
32. T. Rappaport, *Wireless Communications: Principles and Practice*, 2nd ed., Prentice Hall, Upper Saddle River, NJ, 2002.
33. Q. Wang, K. Xu, G. Takahara, and H. Hassanein, Device placement for heterogeneous wireless sensor networks: Minimum cost with lifetime constraints, *IEEE Transactions on Wireless Communications Journal*, vol. 6, no. 7, pp. 2444–2453, 2007.
34. J. Rodrigues, S. Fraiha, H. Gomes, G. Cavalcante, A. De Freitas, and G. De Carvalho, Channel propagation model for mobile network project in densely arborous environments, *Journal of Microwaves and Optoelectronics*, vol. 6, no. 1, pp. 189–206, 2007.

8

PATH PLANNING FOR MOBILE DATA COLLECTORS IN FUTURE CITIES*

Wireless Sensor Networks (WSNs) [1] have evolved from supporting application-specific deployments such as habitat monitoring, health care, and retail supply chains [7], to enabling multiuser platforms that simultaneously support multiple applications operating in a large-scale Internet of Things (IoT) setup [12,25]. Smart cities are examples of such applications that support multiuser access via multi-application platforms. Users might be (1) individuals trying to access information for their personal use, (2) private data centers or public enterprises accessing information periodically to build an information base, and (3) government agencies monitoring information to issue public alerts in case of emergencies. These users may want to access data from any of the following applications supported by the smart city platform: (1) on demand information about availability of free parking spots, (2) periodic information from the city streets and neighborhoods for streets' views and status, and (3) city alerts in case of emergency situations such as major road accidents, which in turn can be used to manage the public transportation traffic. Smart cities are expected to integrate a multitude of wireless platforms and architectures to provide large-scale information access [8,10]. One particularly promising model in this regard is Participatory Sensing (PS) which employs large-scale sensor networks at low cost by utilizing everyday sensory and mobile devices in applications where data is shared among users for the greater public good. In smart cities, under the umbrella of IoT, PS will expand to incorporate heterogeneous data generating/sharing systems including WSNs, database centers, and personal and

---

* This chapter has been coauthored with Mehmet Karakoc and Melih Gunay.

environmental monitoring devices deployed both in the city as well as urban areas. Therefore, we expand the term "sensor" in this chapter to include any form of data source that is either stationary and/or on transit. Hence, this distributed network of sensors will provide a multitude of services to improve the residential experience and quality of living in smart cities. Sensors in such settings are abundant and available with individuals on board private and public vehicles and/or deployed on roads and buildings. In such a comprehensive PS model, an incentive data exchange/delivery approach is required to utilize participants in the sensing process and to ensure data is fairly delivered in a cost effective way. Moreover, the proposed path planning scale introduces challenges regarding the system's limitations in terms of energy consumption, available capacity, and delay. Quality management policies are also to be considered, given the variety of data that is exchanged across the proposed system. Such a smart city environment with multiple users and multiple applications on the same platform is illustrated in Figure 8.1. In this figure, the Base Station (BS) is assumed to be the only central location as a service provider and the assumed smart city contains a number of Access Points (APs). These APs are served with a set of Data Collectors (DCs).

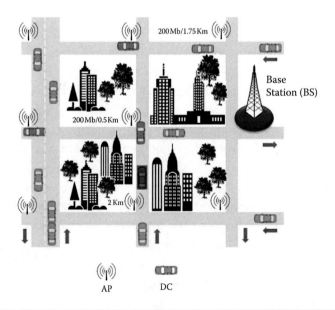

Figure 8.1   Data gathering in a smart city via DCs.

WSNs represent potential candidates for the underlying network implementation in these smart environments because of their inherent ability to observe information from the environment and communicate it wirelessly with end users [11]. However, the ability to handle multiple users simultaneously with different requirements on the data delivered, in terms of attributes such as latency, reliability, and throughput as experienced in an IoT environment, has not received enough research attention. This is primarily because the use of WSNs to enable smart environment applications of the IoT paradigm is still in its experimental stage.

In this chapter, we propose a Hybrid Genetic-based Path Planning (HGPP) approach that minimizes the number of required DCs and their associated traveling lengths while collecting data in a smart city setup. To solve this multi-objective problem, we are using a hybrid meta-heuristic algorithm, the HGPP approach. To this end, our contributions in this chapter can be summarized as follows:

1. We propose a novel data gathering framework for DCs in smart cities, which makes use of mobile vehicles that navigate in a smart environment for efficient data gathering in terms of the number of routes and the total traveling distance.
2. We propose a genetic-based approach, called HGPP, for DCs that incorporate traffic/memory capacity constraints while maximizing the available DCs' utilization in a very competitive time complexity.
3. We compared our approach against a pure genetic algorithm (GA) and a Local Search (LS) algorithm and made recommendations for such kind of approaches in smart city setups.

The remainder of this chapter is organized as follows. Section I reviews related work in the literature. Section II provides the system models and problem description. Section III is the methodology section that provides details about our HGPP approach. Next, our proposed framework is verified and validated in Section IV through extensive simulation results. Finally, we conclude our work in Section V.

8.1 Related work

Recently, WSNs started to render itself as an effective tool in the field of smart cities [10,12]. In such technologies, the sensing process is performed by deploying a number of sensors that are used to collect and send data periodically about the smart city variables (e.g., temperature, humidity, traffic conditions, etc.). The sensor may send the data to a DC via a wireless link. To be able to collect precise data in a timely manner, it might necessitate employing a number of mobile DCs on both public and private transportation vehicles and efficiently utilize their paths of interest.

The DCs are mostly powered by a low energy battery, and hence are expected to have very low transmission power with limited memory and processing capabilities. Consequently, a reliable smart city wireless network will face challenges in terms of connecting with the main BS. Moreover, nonstationary DCs are a must, due to movement of the monitored phenomena (e.g., animals, cars, people, etc.) and limited network resources (e.g., communication range and energy budget). Mobile DCs were proposed in [8,9], assuming centralized knowledge and decisions made at the BS. These mobile nodes move along a set of predefined tracks in the sensing field. It was shown that using DCs (mobile relays) extends the network lifetime compared to conventional WSNs using static sensor nodes only.

To overcome the aforementioned limitations, optimized path planning for DCs is recommended in the applications of smart cities. One of the most adopted methods is the potential field approach [1,2] where attraction (from a targeted location) and repulsion (by obstacles) are used to determine the force to guide the DC to the target. This approach has the drawback of a locally minimal problem. Other methods include a probabilistic road map [3,16], which is based on generating random points and checking for an obstacle collision, or a rapidly expanded random tree [4] where branches of the tree expanded in different directions and are connected to others to generate the path. Other methods include using intelligent systems for path planning [5,6,21]. Although these methods are efficient, incorporating mobile DCs' kinematic constraints using these approaches are not possible in smart city setups where online solutions are a must. Our HGPP approach, on the other hand, can take into consideration DCs and environmental constraints and, as such, have the potential of being used to handle more complex environments with several constraints.

GA is a global search heuristic inspired from Darwin's theory about natural evolution [5,13]. And with the power of this process based on natural selection and genetics, it can be used for solving complex and large combinatorial optimization problems that cannot be solved with conventional methods or old artificial intelligence systems. Unlike the aforementioned DC approaches, GA does not break easily and is robust towards the noise [5]. Each chromosome contains information, and the population means a big data for a generation. GA can be seen as a parallel search approach since it searches for a set of potential solutions in a population. Since GA is a well-used, powerful, and robust optimization technique, it has numerous successful applications. These include path planning, traveling salesman problem, scheduling, Vehicle Routing Problem (VRP), and function optimization to real world problems including image processing, robotics, medicine, biology and more.

GA has a big potential for solving path planning and such problems. In [14], GA was applied by Parvez and Dhar (2013) to effectively find the optimal path while reducing the path cost with shorter computation time and a smaller number of generations in a static environment with known obstacles. Castillo and Trujillo (2005) utilized GA for the problem of offline point-to-point autonomous mobile robot path planning [15]. On the other hand, Achour and Chaalal (2011) used GA to calculate optimized paths in path planning for autonomous mobile robots (e.g., DCs) where a robot is represented by its coaxial coordinates in the targeted environment [17].

In this work, we mainly focused on our hybrid metaheuristic algorithm to improve the search to be applied with environment of unknown obstacles.

8.2 System Models

In this section, we discuss the main system models we assumed in realizing our HGPP approach. We start with the network, followed by the energy and communication models which are considered in this study.

8.2.1 Network Model

In this chapter, we consider a multi-tier WSN framework with three main components: (1) the Base Station (BS), (2) Data Courier (DC), and (3) the WSN Access Point (AP, e.g., cameras) in a smart city.

In a cellular network [18–20,22] there is/are either one BS or multiple BSs. But usually, WSNs have only one BS and this is a typical situation in smart cities. This is the reason why we only consider one BS, which represents the service provider that DCs both start and end with. Different local area networks have different service providers. In this work, we address a local area network (single cell) in order to deliver some data, both for simplicity and without loss of generality.

At the top tier of our proposed architecture, we decide on the order of the targeted intermediate nodes; each has a load to be visited through this BS. A data packet consists of the data traffic that includes loads of a group of APs in the network. Each AP delivers its sensed data by multi-hop transmission through other APs to a mobile DC. DCs are equipped with wireless transceivers and are responsible for forwarding the APs' data loads to the destinations once they are within their communication range.

We assume the following inputs:

1. Each AP has a specific data load to be delivered by a DC.
2. All the DCs in the targeted smart city network are identical.
3. Each DC has a specific storage capacity: If the capacity is as large as the total load of all APs, there should be one route in the solution in which the problem can be formulated like the traveling salesman problem.

We assume the following constraints:

1. Each DC has only one exact route to be utilized.
2. Each route includes a path. The path of any route of a DC includes a subset of the WSN's smart nodes (i.e., APs).
3. Each smart node shall be assigned to exactly one DC (i.e., to be visited only once).
4. The total traffic of a DC (generated by the smart nodes on any route of the DC) cannot be more than the DC storage/BW capacity. However, these DCs can have different speeds and different memory capacities.

We assume the following outputs:

1. The number of required DCs to be used for servicing all the devices in the network.
2. The total length.
3. A set of optimal paths: The routes of a fleet of DCs should be covering all the APs.

Since the path planning problem includes a number of constraints which should be satisfied and aims at minimizing the total cost, it is included in the class of NP-hard problems and requires promising results within a reasonable time. Because of the complexities introduced by various factors, the problem needs heuristics such as GA.

8.2.2 Energy Model

The energy supply of a mobile DC can be unlimited or limited. When a DC has an unconstrained energy supply (rechargeable or simply has enough energy relative to the projected lifetime of the APs), the placement of DCs is to provide connectivity to each AP with the constraint of the limited communication range of APs. When the energy supply of DCs is limited, the allocation of DCs should not only guarantee the connectivity of APs but also ensure that the paths of mobile DCs to the BS are established without violating the energy limitation. In this research, we assume a fixed and limited power supply of DCs.

8.2.3 Communication Model

We consider the exact communication power consumption model used in our previous work [8], where transmitting/receiving power is directly proportional to the total traveling length. We assume that data samples are frequently taken by APs and transmitted wirelessly back to the BS via the mobile DCs based on a synchronized schedule. DCs collect sensed data and coordinate the medium access.

8.3 Hybrid Genetic-based Path Planning (HGPP) Approach

In this section, we explain our HGPP approach. We start with chromosome representation and initial population creation, followed by the path planning.

8.3.1 Chromosome Representation

We represent each candidate solution by one exact chromosome that is a chain of integers where each integer value corresponds to a device or the BS in the network. Each DC identifier uses a separator (a zero value) between a route pair; a string of customer identifiers on a DC's route represents the sequence of a group of deliveries for the DC.

Example 8.1:

Gene sequence: 0—4—5—2—0—8—9—3—1—6—0—7.

   Route#1. BS 4 5 2

   Route#2. BS 8 9 3 1 6

   Route#3. BS 7

For example, in the abovementioned sequence we give a potential solution for the targeted problem with nine devices and three DCs in order to demonstrate the encoding of a chromosome. Note that "0" represents BS, and the number of zeros is equal to the number of required DCs in the solution.

8.3.2 Initial Population Creation

In this step, we choose one of the closest nodes of the intermediate node in forming the path sequence by using the nearest neighbor algorithm [24].

8.3.3 Path Planning

The path planning involves the assignment of the APs in the network to the related DCs and the determination of the visit orders of a group of nodes for each DC. To calculate the total energy of the WSN (E_{total}: sum of the lengths of all routes), the following expression is used:

$$E_{total} = \sum_{r=1}^{R} lengthOfRoute(r) \qquad (8.1)$$

where R is the number of routes in the solution and *lengthOfRoute* is the function that returns the length of rth path.

Fitness function (F for the xth chromosome in the population) of GA is given in Equation (8.2):

$$F(x) = getRouteCount(x) \times \mu + getTotalLength(x) \times \lambda \quad (8.2)$$

Route count function returns the number of required DCs, *total length* function returns the total traveling length. The number of DCs is weighted with value μ (100), and the total length is weighted with value λ (0,001: more sensitive).

Creation of next generation is done in four steps:

1. *Selection* is used for choosing a chromosome pair in order to apply genetic operators.
2. *Crossing-Over* is applied for producing offspring by using genes of selected chromosomes. Parents to mate are selected such that better chromosomes are obtained.
3. *Mutation* is introduced to expand search space using a 2-opt LS algorithm.
4. *Reproduction* is used for the formation of the next generation by using parents and offspring (fitness-based method). Best chromosomes in the current population are directly copied to the next generation. With this approach, called elitism, the best solutions found so far may survive till the end of the run.

Pseudo-code for our HGPP approach is given in Table 8.1.

The time complexity of the HGPP approach is $O(n^2 \log n)$ based on [23]. In LS, neighborhoods of a potential solution are searched for a

Table 8.1   Pseudo-Code of Our HGPP Approach

Function: HGPP()
1.   Create initial population
2.   Improved_LS()
3.   Compute $F(x)$ in Equation (8.2)
4.   **Begin**
5.   Select a chromosome pair by using tournament_selection
6.   Apply crossing-over between the parents
7.   Improved_LS() // on each offspring
8.   Compute $F(x)$ in Equation (8.2)
9.   Create the next generation
10.   **Repeat** (if the progress is not interrupted or max_num_of_generations is not reached)
11.   **Return** best solution in current generation

better outcome. Pseudo-code for our 2-opt LS approach is given in Table 8.2.

For better convergence, LS is applied at the end of each generation. In an offspring when two consecutive zeros appear, one of them is dropped as there is no need to send more than one DC to the same path. This occurs if the representation belongs to a feasible solution, and thus the number of DCs decrease. In path planning, genetic operators could make either inner-route or inter-route modifications. While inner-route modifications change the order of target nodes in a path, inter-route modifications exchange target nodes between paths.

8.4 Performance Evaluation

In this section, we evaluate the performance of the proposed hybrid algorithm by executing our demo 10 times for each experiment and taking the averages of the results.

8.4.1 Experimental Setup

The entire working area of the smart city was partitioned to a set of APs and one BS where a number of DCs, a subset of the moving public transportation vehicles, move between these APs. We assume up to 100 total APs.

8.4.2 Performance Metrics and Parameters

In this research, we evaluate our HGPP approach in terms of four main metrics:

Table 8.2   Pseudo-Code of Our Improved LS Approach

Function: Improved_LS()
1. While (local minimum is not achieved)
2.   Determine the best AP pair: $(i, i+1)$ and $(j, j+1)$
3.   **If** distance$(i, i+1)$ + distance$(j, j+1)$ >
4.   distance(i, j) + distance$(i+1, j+1)$ Then // Improvement...
5.   Exchange the edge
6.   **End If**
7. **End While**

1. *AP count*: The number of devices (nodes not including BS). And each device has a known load.
2. *DC count*: The number of DCs that travel between nodes.
3. *Total traveling length*: The path length of a DC is the sum of the Euclidean distances between nodes that are traveled along the path. Total traveling length on the other hand is the summation of all DCs' path lengths.
4. *Total cost*: The weighted value calculated with the fitness function given in Equation (8.2).

Parameters of our hybrid algorithm are given in Table 8.3.

8.4.3 Simulation Results

In this section, simulation results are shown for various combinations of the data load and the DC capacity under three different approaches: pure GA, pure LS, and different versions of the HGPP approach. These versions vary in the utilized stopping criteria in the proposed HGPP algorithm, which is the maximum number of generations. This value is set to be 200 in HGPP_1, 150 in HGPP_2, and 100 in HGPP_3. Simulation is started with 100 random locations and repeated 10 times with the same location set. Averaged results are plotted in the following figures.

In Figure 8.2, it is observed that total traveled distance is decreasing for all methods (i.e., the *GA*, *LS*, and *HGPP*), while increasing the DC counts. However, the HGPP approach outperforms the GA and LS by at least 50% of their best performance. HGPP_1 approach

Table 8.3 The Parameters of the HGPP Approach

	APPROACH—VALUE		
PARAMETER	LS	GA	HGPP
Population size		100	
Maximum number of generations		100	
Selection method		Tournament	
Crossing-over probability (%)	0%	80%	80%
Mutation probability (%)	5%	0%	5%
Crossing-over operator		Permutation	
LS algorithm		2-opt	
Number of elite chromosomes		2	

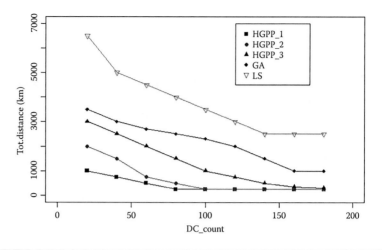

Figure 8.2   Traveled distance vs. the DC count.

achieves the least total traveled distance with respect to all other baseline approaches.

It's worth pointing out here that the HGPP_1, HGPP_2, and HGPP_3 demonstrate the same performance when the total count of DCs is ≥150. Similarly, HGPP_1 and HGPP_2 approaches demonstrate the same performance when the DC count is ≥100. Meanwhile, when the DC count is greater than or equal to 170, all methods cannot improve anymore in terms of the total traveled distance.

In Figure 8.3, as the AP count is increasing, we notice that all approaches, except the GA-based approach, are converging to the same total traveled distance (~1800 km). It's worth pointing out here that the total traveled distance will reach an upper limit that can't be exceeded as long as the total DC count and traffic load are fixed. Unlike the LS and GA approaches, HGPP_1, HGPP_2, and HGPP_3 are monotonically increasing in terms of the total traveled distance as the DC count is increasing. This is because systematic steps have been followed in the proposed HGPP approach towards optimizing the search performance. We also observe that the HGPP_1 has still the lowest total traveled distance, yet it is the fastest approach in terms of convergence. Meanwhile, HGPP_2 and HGPP_3 approaches demonstrate the same performance when the AP count is ≥30. While the AP count is ≥30 all methods cannot improve anymore in terms of the total traveled distance as we noted previously.

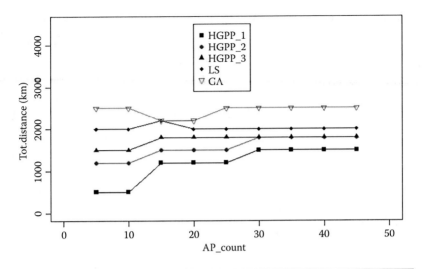

Figure 8.3 Traveled distance vs. the AP count.

In Figure 8.4 we study the AP traffic load effect on the total traveled distances by the occupied DCs. As depicted in Figure 8.4, the total traveled distance is increasing for all approaches, except the HGPP_1, while increasing the AP load from 10 to 80 MB. This indicates a great advantage for the proposed HGPP approach while using a bigger max_generation value, where no more traveled distance is

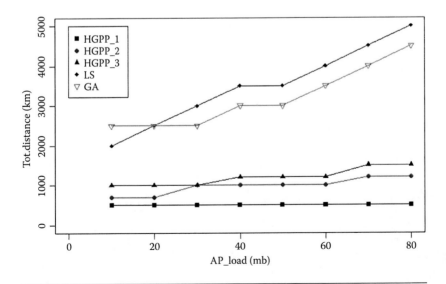

Figure 8.4 Traveled distance vs. the AP load.

required for a few more extra MBs at the APs. It's obvious as well how the LS and GA approaches are monotonically increasing with a larger slope than HGPP_2 and HGPP_3, which means that LS and GA are more sensitive to the AP loads to be delivered. This can be a significant drawback in the IoT and a big data error. Obviously the HGPP_1 experiences the lowest total traveled distance while the AP load is increasing. Also, we notice that when the AP load is ≥70 MB, the HGPP_2 and HGPP_3 can't improve anymore in terms of the total traveled distance.

On the other hand, the total traveled distance is decreasing for all experimented approaches while the DC capacity is increasing as shown in Figure 8.5. In this Figure, HGPP_1, HGPP_2, and HGPP_3 demonstrate similar performance when the DC capacity is ≥60. HGPP_1, HGPP_2, and HGPP_3 are very close to each other in terms of the total traveled distance as the DC capacity increases. And again HGPP_1 outperforms the other approaches in terms of the total distance. On the other hand, GA and LS are the worst in terms of the total traveled distance as the DC capacity increases.

In Figure 8.6 we study the AP count effect on the total required DCs. Obviously, the DC count is increasing monotonically for all approaches while the AP count is increasing. Surprisingly, the LS and

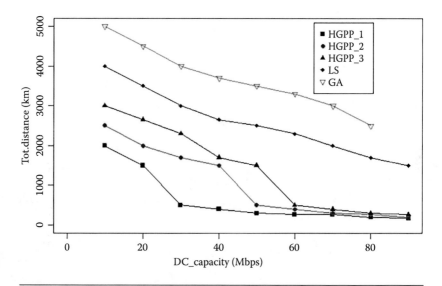

Figure 8.5   Traveled distance vs. the DC capacity.

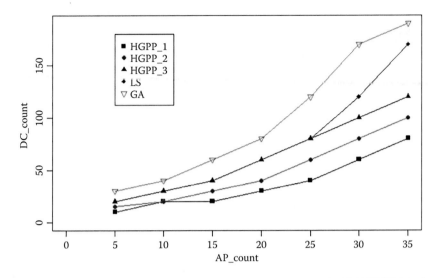

Figure 8.6 Required DC count vs. the AP count.

HGPP_3 approaches demonstrate the same performance when the AP count is ≤25. This can be returned for the small number of generations that has been used with the HGPP_3 approach. Meanwhile, HGPP_1 necessitates the lowest DC count while increasing the AP count. And GA necessitates the most DC count while the AP count is increasing.

In Figure 8.7, the DC count is increasing monotonically for all methods while the deployment area size is increasing. The GA and LS approaches demonstrate the same performance when the DC count is ≥3000 km² and these approaches necessitate the largest DC count while the area size is increasing. Meanwhile, HGPP_1 necessitates the lowest DC count as the area size is increasing.

In Figure 8.8, we study the cost effect on the average system throughput measured in Mbps. As depicted by Figure 8.8, the overall average system throughput is increasing for all methods while the cost parameter is increasing. However, they saturate after reaching a specific cost value (~1000) in terms of their overall achieved throughput. For example, HGPP_3 is saturating at a cost equal to 1000 and its corresponding throughput is not enhancing anymore once it reaches 127 Mbps, regardless of the cost amount. HGPP_2 and HGPP_3 demonstrate the same performance once the system cost reaches a specific value (~1000). On the other hand, the GA and LS approaches

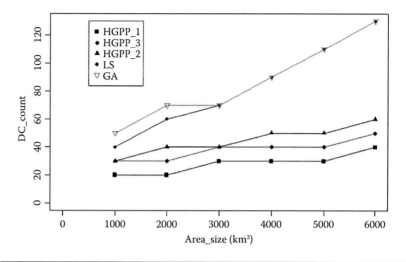

Figure 8.7 Required DC count vs. the targeted region size.

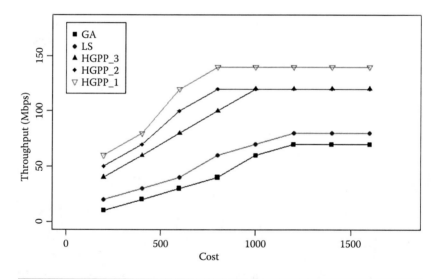

Figure 8.8 Throughput vs. the cost.

demonstrate almost the same performance over all cost values. However, HGPP_1 outperforms all approaches under all examined cost values. The GA and LS have the lowest throughput.

Further experimental results are detailed in Table 8.4 for more comprehensive analysis about the targeted three approaches, GA, LS, and HGPP, in this study. Results in Table 8.4 confirm that total traveled distances increase monotonically while increasing the number of DCs. The worst increase in total traveled distances is observed while applying the pure GA approach as opposed to the proposed HGPP approach. In addition, the proposed hybrid approach converges much faster than other heuristic approaches (e.g., LS and pure GA) with the least number of generations.

Results in Table 8.4 confirm the minimum cost achievement while applying the proposed hybrid approach HGPP. And the optimal set of paths is found within a competitive time complexity as opposed to the other two approaches. Consequently, the proposed approach in this research clearly outperforms other classical heuristic approaches such as the ones that depend heavily on LS algorithms and the ones which purely apply GA in terms of the found solutions as well as the overall time complexity.

8.5 Conclusions

In this chapter we introduce HGPP, a hybrid IoT public sensing framework for smart cities. Our framework is based on a multitier architecture that caters for heterogeneous data sources (e.g., sensors) in addition to mobile data collectors in urban areas which are isolated from their corresponding data processing center (BS). According to our framework, access points at the top tier of the proposed architecture receive sensor readings and initiate delivery requests. Our delivery scheme implements genetic-based algorithms that realize distance and cost sensitivities. At the top tier, our cost-based model employs a fitness function that maximizes the network operator gain according to the limited DCs count, storage capacity, and total traveled distances. We provide simulation results showing the efficiency of our framework when compared to two prominent heuristic approaches. Our simulation results show that the HGPP framework exhibits superior performance for different network sizes, storage capacities, DCs' counts, and end-to-end traveled distances.

Our future work will investigate the utilization of the same DCs in smart city settings with non-deterministic mobility trajectories. It will also investigate the same problem while considering dynamic APs in the vicinity of the employed DCs.

Table 8.4 Numerical Results for the Proposed HGPP in Comparison to LS and GA

TEST			APS VS. THE TOTAL TRAVELED DISTANCES						
APs Count			10	25	40	55	70	85	100
Routes' count (DCs to be used)			1	1	2	3	3	4	4
APPROACH									
GA	Length	Avg.	1956.7	4462.9	7651.2	11292.4	14189.1	17778.9	21636.2
		Std. dev.	31.02	302.48	334.92	531.53	529.31	745.98	967.65
	Generation	Avg.	55.8	89.8	87.9	89	94.6	96	95.5
		Std. dev.	18.09	8.62	7.44	7.64	6.1	4	3.69
	Time	Avg.	188	392.2	620.3	847.9	1206	1945.5	1970.9
		Std. dev.	52.04	43.36	66.27	98.82	76.36	116.07	204.95
LS	Length	Avg.	1901	2315	3209.3	3923.1	4416.6	5057.4	5530.2
		Std. dev.	0	0	24.73	47.8	53.26	44.29	121.27
	Generation	Avg.	1.4	2.9	5.5	5.1	3.3	4.3	3.8
		Std. dev.	0.92	2.66	2.29	4.37	2.05	2.79	2.32
	Time	Avg.	3.1	19	76.9	118.7	195.9	347.8	472.6
		Std. dev.	6.24	17.16	24.95	74.96	91.76	185.52	241.47
HGPP	Length	Avg.	1901	2315	3172	3820.7	4252.4	4897.2	5243.8
		Std. dev.	0	0	0	32.04	21.93	18.8	48.25
	Generation	Avg.	1.5	1.4	17.8	76.7	85.1	65.5	87.7
		Std. dev.	0.92	0.92	13.97	21.66	18.2	22.98	12.39
	Time	Avg.	2.7	7	21.6	145.7	168	257	353.7
		Std. dev.	6.29	8.32	141.85	386.95	491.89	1193.37	915.8

References

1. G. Singh, and F. Al-Turjman, A data delivery framework for cognitive information-centric sensor networks in smart outdoor monitoring, *Elsevier Computer Communications Journal*, vol. 74, no. 1, pp. 38–51, 2016.
2. A. A. Masoud, A harmonic potential field approach for joint planning and control of a rigid, separable nonholonomic, mobile robot, *Elsevier: Robotics and Autonomous Systems*, vol. 61, no. 6, pp. 593–615, 2013.
3. Y. Zhang, N. Fattahi, and W. Li, Probabilistic roadmap with self-learning for path planning of a mobile robot in a dynamic and unstructured environment, In *IEEE International Conference on Mechatronics and Automation (ICMA)*, Takamatsu, Japan, August 2013.
4. S. M. LaValle and J. J. Kuffner, Rapidly-exploring random trees: Progress and prospects, 2000. http://citeseerx.ist.psu.edu/viewdoc/summary?doi=10.1.1.38.1387.
5. J. Tiu and S. X. Yang, Genetic algorithm based path planning for mobile robots, In *IEEE Conference on Robotics and Automation (ICRA)*, Taipei, Taiwan, September 2003.
6. L. Wang, S. X. Yang, and M. Biglarbegian, A fuzzy logic based bio-inspired system for mobile robot navigation, In *IEEE Conference on Multisensor Fusion and Integration for Intelligent Systems (MFI)*, Hamburg, Germany, September 2012.
7. F. Al-Turjman, H. Hassanein, and M. Ibnkahla, Efficient deployment of wireless sensor networks targeting environment monitoring applications, *Elsevier Computer Communications Journal*, vol. 36, no. 2, pp. 135–148, 2013.
8. W. Alsalih, H. Hassanein, and S. Akl, Routing to a mobile data collector on a predefined trajectory, In *Proceedings of IEEE International Conference on Communications (ICC)*, Dresden, Germany, 2009, pp. 1–5.
9. A. Azad and A. Chockalingam, Mobile base stations placement and energy aware routing in wireless sensor networks, In *Proceedings of IEEE Wireless Communications and Networking Conference (WCNC)*, Las Vegas, NV, 2006, pp. 264–269.
10. A. Al-Fagih, F. Al-Turjman, W. Alsalih, and H. Hassanein, A priced public sensing framework for heterogeneous IoT architectures, *IEEE Transactions on Emerging Topics in Computing*, vol. 1, no. 1, pp. 133–147, 2013.
11. F. Al-Turjman, H. Hassanein, W. Alsalih, and M. Ibnkahla, Optimized relay placement for wireless sensor networks federation in environmental applications, *Wiley: Wireless Communication and Mobile Computing Journal*, vol. 11, no. 12, pp. 1677–1688, 2011.
12. F. Al-Turjman, H. Hassanein, and S. Oteafy, Towards augmenting federated wireless sensor networks, In *Proceedings of the IEEE International Conference on Ambient Systems, Networks and Technologies (ANT)*, Niagara, ON, 2011, pp. 224–231.

13. S. Al-Harbi, F. Noor, and F. Al-Turjman, March DSS: A new diagnostic march test for all memory simple static faults, *IEEE Transactions on CAD of Integrated Circuits and Systems*, vol. 26, no. 9, pp. 1713–1720, 2007.

14. W. Parvez and S. Dhar, Path planning of robot in static environment using genetic algorithm (GA) technique, *International Journal of Advances in Engineering and Technology*, vol. 6, no. 3, p. 1205, 2013.

15. O. Castillo and L. Trujillo, Multiple objective optimization genetic algorithms for path planning in autonomous mobile robots, *International Journal of Computers, Systems and Signals*, vol. 6, no. 1, pp. 48–63, 2005.

16. F. Al-Turjman, H. Hassanein, and M. Ibnkahla, Quantifying connectivity in wireless sensor networks with grid-based deployments, *Elsevier: Journal of Network and Computer Applications*, vol. 36, no. 1, pp. 368–377, 2013.

17. C. E. Thomaz, M. A. C. Pacheco, and M. M. B. R. Vellasco, Mobile robot path planning using genetic algorithms, In Mira, J., Sánchez-Andrés, J. V. (eds) *Foundations and Tools for Neural Modeling. IWANN 1999.* Lecture Notes in Computer Science, vol. 1606, Springer, Berlin and Heidelberg, 1999.

18. ITU-T Series Y recommendation: ITU-T Y.2221; Requirements for support of ubiquitous sensor network applications and services in the NGN environment, January 2010.

19. Qualcomm, LTE-advanced: Heterogeneous networks. www.qualcomm.com/media/documents/lte-heterogeneous-networks, 2011.

20. I. Ahmed, B. Qazi, and J. Elmirghani, Base stations locations optimisation in an airport environment using genetic algorithms, In *Proceedings of the International Wireless Communications and Mobile Computing Conference (IWCMC)*, 2012, pp. 24–29.

21. F. Al-Turjman, H. Hassanein, and M. Ibnkahla, Towards prolonged lifetime for deployed WSNs in outdoor environment monitoring, *Elsevier Ad Hoc Networks Journal*, vol. 24, no. A, pp. 172–185, 2015.

22. S. Strzyz, K. I. Pedersen, J. Lachowski, and F. Frederiksen, Performance optimization of pico node deployment in LTE macro cells, In *Proceedings of the Future Network Mobile Summit (FutureNetw)*, 2011, pp. 1–9.

23. G. Singh and F. Al-Turjman, Learning data delivery paths in QoI-aware information-centric sensor networks, *IEEE Internet of Things Journal*, vol. 3, no. 4, pp. 572–580, 2016.

24. B. A. AlSalibi, M. B. Jelodar, and I. Venkat, A comparative study between the nearest neighbor and genetic algorithms: A revisit to the traveling salesman problem, *International Journal of Computer Science and Electronics Eng. (IJCSEE)*, vol. 1, no. 1, pp. 34–38, 2013.

25. F. Al-Turjman, Cognition in information-centric sensor networks for IoT applications: An overview, *Annals of Telecommunications*, pp. 1–16, doi:10.1007/s12243-016-0533-8, 2016.

9

CONCLUSIONS AND FUTURE DIRECTIONS

The last decade has witnessed a growing interest of Outdoor Environment Monitoring (OEM) applications in Wireless Sensor Networks (WSNs) due to their unique potential in the remote detection and prevention of disasters. However, such a critical mission requires extraordinary efficiency to ensure data availability and timely delivery within a reasonable cost. Thus, the deployed WSNs' connectivity and lifetime along with fault-tolerance and cost-effectiveness are key network properties in OEM.

In this book, we argue for exploiting heterogeneous (sensors/relays) and/or hybrid (mobile/static) nodes on generic 3-D grid models to satisfy OEM-specific network properties in practice. Existing proposals (schemes) in this direction are limited to special cases of the problem and/or provide results that can be arbitrarily far from the optimal ones. In addition, the majority of these proposals are not considering practical issues in OEM applications, including placement uncertainty, communication irregularity, and node redundancy/mobility. Therefore, we investigated these issues while considering 3-D grid-based deployment in Chapter 3 and proposed other promising deployment schemes in Chapters 4 through 8.

In fact, significance of the proposed deployment schemes in this book moves beyond the scientific data collection to enabling an intelligent and safe living environment. These deployment schemes provide reliable interaction with the network users anytime and anywhere. It plans the deployment of sensor/relay nodes in transportation systems to provide real time traffic information for users inside their moving vehicles. Furthermore, it optimizes the node deployment in forests to detect fires and report wildlife activities and in water bodies to record events pertaining to floods, water pollution, coral reef conditions, and oil spills. It also targets other rural and hazardous areas such as deserts, polar and volcanic terrains, and battlefields.

Investigated grid practicality and presented deployment schemes in this book are summarized in Section 9.1. Some future research directions are outlined in Section 9.2.

9.1 Summary

In Chapter 3, we introduced a novel 3-D deployment strategy, called Optimized 3-D deployment with Lifetime Constraint (O3DwLC), for relay nodes in environmental applications. The strategy optimizes network connectivity while guaranteeing specific network lifetime and limited cost. The effectiveness of this strategy is validated through extensive simulations and comparisons, assuming practical considerations of signal propagation and connectivity. In Chapter 4, we proposed two Optimized Relay Placement (ORP) strategies with the objective of federating disjointed WSN sectors with the maximum connectivity under a cost constraint on the total number of RNs deployed. The performance of the proposed approach is validated and assessed through extensive simulations while assuming practical considerations in outdoor environments. In Chapter 5, we present a novel communication protocol, Fixing Augmented network Damage Intelligently (FADI), based on a minimum spanning tree construction to reconnect the disjointed WSN sectors. This protocol outperforms the related work in terms of the average relay node count and distribution, the scalability of the federated WSNs in large scale applications, and the robustness of the topologies formed. We elaborate further on this strategy in Chapter 6, while the hexagonal virtual grid is assumed instead of the square one. We propose the use of cognitive nodes (CNs) in the underlying sensor network to provide intelligent information processing and knowledge-based services to the end users. Chapter 7 proposes a 3-D grid-based deployment for heterogeneous WSNs (consisting of (SNs), static RNs, and mobile RNs). The problem is cast as a Mixed Integer Linear Program (MILP) optimization problem with the objective of maximizing the network lifetime while maintaining certain levels of fault-tolerance and cost-efficiency. Moreover, an Upper Bound (UB) on the deployed WSN lifetime, given that there are no unexpected node/link failures, has been driven. A typical scenario has been discussed and analyzed, as well, in smart cities while considering genetic based approaches in Chapter 8. In this chapter, we study the path planning problem for these data collectors (DCs) while optimizing their counts and their total traveled distances. As the total collected load on a given DC route cannot exceed its storage capacity, it is important to decide on the size of the exchanged data packets (images, videos, etc.) and the sequence of the targeted data sources to be visited. We propose a hybrid heuristic approach for public data delivery in smart city settings. Extensive simulations are performed; the results confirm the effectiveness of the proposed approach in comparison to other heuristic approaches with respect to total traveled distances and overall time complexity.

9.2 Future Work

So far, several future research directions and open issues can be derived from our work. In this section, we outline some of these directions.

1. In Chapter 3, we examined the resilience of grid-based deployments when placement uncertainty and communication irregularity are assumed. Future work could investigate the resilience of the grid connectivity under other permanently hindering conditions, such as node failures that could occur due to physical damage, limited energy, or temporarily hindering conditions like mobile obstacles (e.g., wild animals).
2. Investigating the properties of grid deployment under varying node transmission ranges and/or energy levels is a promising direction as well.
3. In Chapter 5, we provide a k fault-tolerant WSN; however, it would be of great interest to also find the optimal k value, given specific PNF and PDN values. We expect the optimal value to vary in various situations depending on the environment harshness.
4. Also, we intend to expand our work in a more dynamic WSN where each sensor node has the mobility feature and to check the effect on the network fault-tolerance efficiency.

5. In Chapter 7, we provide a guaranteed lifetime period equal to LT by maximizing the network connectivity. Also of practical interest would be investigating the optimal value of LT that overcomes specific PNF and PDN.

6. In this paper, we introduce HGPP—a hybrid Internet of Things (IoT) public sensing framework for smart cities. Our future work would investigate the utilization of the same DCs in smart city settings with nondeterministic mobility trajectories.

7. It will also investigate the same problem while considering dynamic access points (APs) in the vicinity of the employed DCs.

Index